W0037271

Progressive Energy Policy

Series Editors
Caroline Kuzemko
The University of Warwick
Coventry, UK

Catherine Mitchell
University of Exeter
Penryn, UK

Andreas Goldthau
Royal Holloway University of London
Egham, UK

Alain Nadaï
International Research Centre on Environment and
Development (CIRED-CNRS)
Nogent-sur-Marne, France

Shunsuke Managi
Kyushu University
Fukuoka, Japan

Progressive Energy Policy is a new series that seeks to be pivotal in nature and improve our understanding of the role of energy policy within processes of sustainable, secure and equitable energy transformations. The series brings together authors from a variety of academic disciplines, as well as geographic locations, to reveal in greater detail the complexities and possibilities of governing for change in energy systems. Each title in this series will communicate to academic as well as policymaking audiences key research findings designed to develop understandings of energy transformations but also about the role of policy in facilitating and supporting innovative change. Individual titles will often be theoretically informed but will always be firmly evidence-based seeking to link theory and policy to outcomes and changing practices. Progressive Energy Policy is focussed on whole energy systems not stand alone issues; inter-connections within and between systems; and on analyses that moves beyond description to evaluate and unpack energy governance systems and decisions.

More information about this series at
http://www.palgrave.com/gp/series/15052

Béatrice Cointe • Alain Nadaï

Feed-in tariffs in the European Union

Renewable energy policy, the internal electricity market and economic expertise

Béatrice Cointe
TIK Centre for Technology,
Innovation and Culture, University of
Oslo
Oslo, Norway

Alain Nadaï
CIRED-CNRS (Centre International
de Recherche sur l'Environnement et
le Développement)
Nogent-sur-Marne, France

Progressive Energy Policy
ISBN 978-3-030-09464-5   ISBN 978-3-319-76321-7   (eBook)
https://doi.org/10.1007/978-3-319-76321-7

ACKNOWLEDGEMENTS

This book started as an enquiry into the origins of feed-in tariffs. This interest was raised by a study of solar energy policy in France, carried out in the context of a PhD thesis that was written between 2010 and 2014 at the Centre for International Research on Environment and Development (CIRED). The first version of this account was a chapter of the aforementioned PhD thesis, which has been revised, extended, and updated. This work benefitted from financial support from the DIM R2DS Ile-de-France under Grant no. 2010-13 and from the French National Research Agency (ANR) under Grant no. 2011-SOIN-003-01 (COLLENER Project). The study has been presented in several contexts over the years, and we are indebted to comments from colleagues at CIRED, from members of the PhD panel, and from discussants in conferences. We would like to thank in particular Fabian Muniesa, Noortje Marres, Peter Karnøe, Thomas Reverdy, and Philippe Quirion for helpful insights and suggestions at different stages of our research. All remaining errors of course remain entirely our own.

Contents

CONTENTS

LIST OF ABBREVIATIONS

ANT	Actor-Network Theory
DG Competition	European Commission Directorate General for Competition
DG Research	European Commission Directorate General for Research
EC	European Communities
EEG	*Erneuerbare-Energien Gesetz* (Renewable Energy Law)
EU	European Union
EU-ETS	European Union Emission Trading Scheme
FIP	Feed-in Premium
FIT	Feed-in Tariff
IPCC	Intergovernmental Panel on Climate Change
NFFO	Non-Fossil Fuel Obligation
PURPA	Public Utility Regulatory Policies Act
PV	Photovoltaics
RE	Renewable Energy
RES-E	Electricity from Renewable Energy Sources
RET	Renewable Energy Technologies
SET Plan	Strategic Energy Technologies Plan
SRREN	Special Report on Renewable Energy
StrEG	*Stromeinspeisungsgesetz* (Feed-in Law)
STS	Science and Technology Studies
TGC	Tradable Green Certificate
VDEW	*Verband der Elektrizitätswirtschaft*

LIST OF TABLES

CHAPTER 1

Agencing Feed-in Tariffs in the European Union

Abstract The introductory chapter describes the intention of the book and provides an overview of feed-in tariffs and renewable energy policy in the European Union (EU). It outlines the perimeter and analytical approach of the book. Cointe and Nadaï first describe feed-in tariffs and their origins. They review milestones of European renewable energy policy and their relations to the diffusion of feed-in tariffs in Member States. Having provided this background information, Cointe and Nadaï account for their choice to rely on a combination of documentation from European institutions, expert sources, and academics. They define socio-technical agencements and explain what an analysis of feed-in tariffs as agencements brings to an understanding of renewable energy policy and EU policy.

Keywords Feed-in tariffs • European Union • Agencement • Performativity • Liberalisation

Throughout Europe, renewable energy production has expanded over the past decades, boosted by support policies of various kinds. The European Union (EU) appears as a driving force in the deployment of renewable energy and renewable energy support policies, especially for electricity generation. Since the late 1990s, it has encouraged Member States to develop renewables; adopted directives setting mid-term targets for this

B. Cointe, A. Nadaï, *Feed-in tariffs in the European Union*, Progressive Energy Policy, https://doi.org/10.1007/978-3-319-76321-7_1

development; and produced regular reports assessing progress and evaluating the merits of the policies introduced in Member States.

The EU's renewable energy policy emerged and developed along with the liberalisation of the electricity market, which was officially launched by the 1996 Directive on the internal electricity market. The process of integrating the European electricity market and the process of integrating electricity from renewable energy sources in this electricity market unfolded more or less at the same time, but not necessarily in perfect tune—especially since Member States' perspectives and objectives in introducing support for electricity from renewable energy sources (RES-E) were not always aligned with those of the European Commission. Simultaneously, renewable energy policy consolidated as a field of research and expertise, attracting increasing academic attention. This book investigates this process from the vantage point of the history and evolutions of one kind of RES-E policy instruments, feed-in tariffs (FITs).

While FITs have been a dominant form of support for RES-E in European countries, their relationship to the policy principles of the EU has always been rather ambiguous: the European Commission has gone back and forth in granting them the label of "market-based instruments" (which amounts to a validation from its own very market-oriented perspective), and has alternated between shunning them and recognising them as the most effective form of support for RES-E. At any rate, ever since renewable energy made its entrance within the scope of European concerns, FITs have been part of the picture, and their vices and virtues have been debated.

The research that led to this book started from an ambition to understand where FITs came from. In retracing their origins and genealogy, we soon encountered European institutions and legislation. We also found a large body of academic and grey literature investigating the characteristics, design, histories, and effects of FITs, often in comparison with other instruments. From there emerged the project to retrace the European career of FITs on the basis of what the EU as well as the literature had said about them. In this book, we thus follow FITs in some of the countries that have implemented them and through the political and academic debates about EU electricity policies and markets, from the late 1970s to 2015. The picture we draw is certainly not exhaustive given the short format in which it is displayed, but hopefully provides an overview of the trajectory of FITs and of the various concerns and issues that have been attached to them over the years.

This book then tells the story of an instrument of renewable energy policy, but it also considers this story as a vantage point from which to look into the wider evolutions, tensions, and frictions at play in European renewable energy policy. In particular, it gives insights about the objectives of liberalisation and harmonisation that have been at the heart of the European project for some decades now, and about the priorities that have guided them in the case of the electricity sector (Barry 1993; Doganova and Laurent 2016). The debates that surrounded the evolutions of FITs, and the evolution of FITs in Member States itself, also interrogate the notion that a balanced, well-functioning, liberalised market can serve as a device to serve the common good and solve problems that are not reduced to economic and market activity (Geiger et al. 2014; Doganova and Laurent 2016). Even if renewable energy policy is a small part of the internal electricity market project, and FITs are only one element of renewable energy policy, we argue that a close look at FITs can shed light on these issues in several ways.

First, it leads us to look into the Commission's perspective on liberalisation and harmonisation: the liberalisation and harmonisation agendas run through the Commission's discourses on renewable energy policy. The ideal model of a smoothly running internal market for electricity largely shapes the Commission's conceptions and assessments of RES-E policy intervention, so in this case we will watch it deploy in relations to a specific issue. Second, it gives us insight into the actual unfolding of the liberalisation and harmonisation projects, and into their imperfect realisation. In our study, we see how actual RES-E policy development in member states sometimes clashed with the Commission's ideals and ambitions, and how such tensions have been resolved.

We look at how these two enactments of European renewable energy policy—at the EU level and Member States—play out in the design and management of one particular type of instruments and in the debates it raised. FITs are a particularly intriguing object in this respect. The Commission has for a long time considered them to go against the internal market project; but for a few years, between 2005 and 2011, it dubbed them as "generally the most efficient and effective support schemes for promoting renewable energy" (Commission of the European Communities 2008). We thus investigate to what extent different designs and conception of FITs have contributed to the integration of both RES-E and diverse national energy policy agendas (partly informed by economic or industrial interests) within the projected EU internal electricity market, as well as within the EU's overarching energy-climate and innovation agendas.

In retracing the trajectory of FITs, we pay particular attention to two aspects of it. First, we are interested in the simultaneous production of guiding policy principles, actual policies, and expertise on these policies (and on their potential to align to guiding principles). This implies that we look at the interweaving of theoretical and practical concerns in the evolution and evaluation of FITs in Europe. Second, we are attentive to the relative importance of environmental objectives and liberalisation objectives as they have influenced the design and the theorisation of FITs. This translates into an attention to the extent to which FITs are described, and constituted, as "market-based" instruments. It also leads us to interrogate what lies behind the term "market-based": we follow evolutions in this conception, noting shifts in focus from competition to investment, which draw attention to different characteristics of market activities and different "virtues" of so-called market-based policies.

This first chapter sets the scene for our study. We start by providing background on feed-in tariffs: How do they work? When did they appear? How do they relate to EU energy policy? After that, we explain how we approached them. We describe the material we relied upon and detail our strategy to explore broader trends and tensions in EU energy policy from the study of the trajectory of one type of instrument. In so doing, we relate our study to current sociological work, especially in science and technology studies (STS) and economic sociology, highlighting how we can contribute to ongoing debates. The book then follows a chronology punctuated by key EU directives relevant to RES-E policy. Each period (1970s–1996; 1996–2000; 2001–2008; 2008–2015) provides an opportunity to explore different aspects of the making of FITs and of their relationship with EU policy.

FEED-IN TARIFFS: AN OVERVIEW

Defining Features and Brief History

Feed-in tariffs are commonly defined as state-backed incentives to invest in the generation of electricity from renewable sources. They organise access to the electricity grid and markets for RES-E, and ensure a stable return on investment in renewable energy technologies. To do so, they usually combine three elements: a purchase obligation, implying that utilities must purchase electricity produced from renewable sources and feed it to the grid; a fixed tariff for the purchase of this electricity, the level of which is determined by the regulator; and a fixed period over which the said tariff is guaranteed, usually reflecting the lifetime of an installation (Jacobs 2010, p. 287; Couture and Gagnon 2010). They usually involve mechanisms to compensate for the extra costs induced for utilities, often in the form of a levy on electricity use.

FITs thus rely on a simple principle: by guaranteeing the financial viability of RES-E generation and organising its integration in the electricity system, they provide an "almost risk-free contract" from the perspective of renewable energy producers (Mitchell et al. 2011, p. 50). They are meant to be transitory: they should drive and accelerate the uptake of still expensive energy technologies, up to the point that these technologies reach market competitiveness and no longer need support. They are usually classified as "price-based" instruments, in that they set a required price for renewable electricity, but not a desired quantity—as opposed to quota systems such as the US Renewable Portfolio Standards or the Tradable Green Certificates adopted in some European countries in the 2000s. The actual development of FITs, however, paints a more complex picture.

In practice, FITs come in multiple forms, and in diverse degrees of sophistication. As the IPCC Special Report on Renewable Energy stated, "FITs can be very simple [...] or they can be quite complex" (Mitchell et al. 2011, p. 52). Indeed, as each of their features is negotiated politically—from the definition of what qualifies as electricity from renewable energy sources to the level of remuneration and the modalities for compensating utilities, including contract duration and grid-connection procedures—FITs are shaped by political priorities, country-specific balance of powers in energy policy, and a range of context-dependent considerations.

FITs find their origin in the late 1970s. In the US, the 1978 Public Utility Regulatory Policies Act (PURPA) introduced an obligation for electric utilities to purchase the electricity generated by small-scale producers, at prices reflecting avoided costs for utilities (Loiter and Norberg-Bohm 1999; Lesser and Su 2008). In Europe, similar mechanisms first appeared in Denmark and Germany in 1979, in the form of voluntary agreements between electric utilities and wind power producers, chiefly to integrate growing wind power capacities into the electricity system. They were introduced into German and Danish legislations in the early 1990s (1991 in Germany, 1992 in Denmark), with Spain following a few years later. However, they only became a widespread instrument for the support of renewable electricity in the 2000s.

The German *Erneuerbare-Energien Gesetz* (EEG), adopted in 2000, was particularly influential. By setting technology-specific FITs and planning for the decrease of feed-in rates at a rhythm following expected evolutions in technology costs, the EEG model integrated considerations about renewable energy technologies and their cost dynamics. It thus turned FITs into innovation-oriented instruments designed to steer the deployment of RES-E installations. Its revision in 2004 accentuated this dimension. FITs were subsequently adopted in a growing number of countries (among which France, Czech Republic, Italy, Portugal, Greece, Austria, Belgium, Estonia, Hungary, and later, the UK), to the extent that "by early 2010, at least 45 countries had FITS at the national level (including much of Europe)" (Mitchell et al. 2011, p. 14).

Over the course of their diffusion, FITs were adapted and sophisticated. From country to country, but also in their evolution in one specific country, they have varied in scope, rationale, design, and effects. They were tailored to specific technologies and made to incorporate several distinct objectives, from contributing to greenhouse gas emission reductions to supporting industries. They also had to be reformed to address issues triggered by their own effects, with more or less success. FITs often spurred rapid increases in RES-E capacity, generating high collective costs as well as difficulties with grid management or electricity spot markets, and various strategies have been used to address these problems (Karnøe 2013; Hoppmann et al. 2014; Cointe 2015, 2017).

Why Study FITs in the European Union?

The adoption of FITs in several EU Member States was not directly driven by the European Union. Indeed, the relationship of FITs with EU policy principles has always been ambiguous. The European Commission initially tended to disapprove of such instruments, considering them as state intervention in the operation of markets and as potential distortions to competition. All the same, the promotion of renewable energy in Member States was framed by EU directives, and in that sense feed-in tariffs were part of EU renewable energy policy. Their emergence, diffusion and evolution also occurred in the context of the liberalisation of the EU electricity market. The trajectory of FITs thus needs to be understood in relation to these two dimensions of EU energy policy (namely, the promotion of renewable energy and liberalisation), as well as to the interplay between EU policies and national policies.

By all accounts, the liberalisation agenda has been a building block of EU energy policy over the past three decades. The liberalisation of the energy sector, along with that of telecommunications and financial markets, was initiated by the European Commission in the late 1980s, in the hope of reviving European integration and the Internal market programme (Jabko 2006; Reverdy 2014). This was part of broader institutional transformations oriented by market integration as the backbone of European integration (Eising 2001). Liberalisation implies a diminution of direct state intervention in the organisation and regulation of markets, on the basis of a liberal critique of state action (viewed as prone to capture and unpredictable, thereby undermining investors' confidence). Market regulation becomes entrusted to independent regulatory authorities and guided by economic expertise (Reverdy 2014, p. 55).

As far as electricity is concerned, the liberalisation agenda materialised in the 1996 Directive on the internal electricity market (European Parliament and Council 1996). Following extensive debates, the directive was relatively ambiguous, and left margins for manoeuvre for Member States: as Reverdy noted, it followed a "logic of experimentation", insofar as it did not seek to "set the organisation of the market once and for all, but to engage in a process of gradual exploration and adjustment" (Reverdy 2014, p. 60, authors' translation).

Over the same period, that is from the late 1980s onwards, European institutions began to take interest in renewable energy, starting with a Council Recommendation "on developing the exploitation of renewable energy sources" in 1988 (European Council 1988). The Commission worked on a strategy for renewable energy throughout the 1990s (Commission of the European Communities 1996, 1997, 1999). This led to a first Directive "on the promotion of electricity produced from renewable energy sources in the internal electricity market", which set indicative targets for the contribution of RES-E to gross electricity consumption in Member States, with an EU-wide objective of 22% of gross electricity consumption produced from renewable energy sources by 2010 (European Parliament and Council 2001). In 2008, The Energy-Climate Package took it further with objectives to 2020, explicitly drawing European energy and climate policies together. As part of this package, the Directive "on the promotion of the use of energy from renewable sources" set binding national targets for Member States (European Parliament and Council 2009). During the preparation of both the 2001 and the 2009 directives, the harmonisation of renewable energy support schemes throughout the

EU was on the agenda, with the Commission supporting a European Tradable Green Certificates (TGC) scheme, but it did not make it through either directive. The Commission regularly reviewed renewable energy policies in Member States and published several "renewable energy progress reports" (Commission of the European Communities 2004; European Commission 2011, 2013). In 2014, it issued "Guidelines on State aid for environmental protection and energy", advocating a gradual phase-out of FITs and their replacement by either market premiums or tendering schemes (European Commission 2014). In 2015, the Energy Union Package was adopted; articulated around five dimensions, it includes an EU target for the share of renewable energy in energy consumption in 2030, albeit not a very ambitious one (27% of renewable energy in total energy consumption, to be compared to the 20% by 2020 target) (European Commission 2015).

The gradual elaboration of an EU strategy for renewable energy policy constituted a framework and acted as a driver for the setting-up of renewable energy policies in individual Member States, even if the choice and design of support instruments was left to them. All this occurred in the context of the liberalisation of the electricity sector, and while renewable energy policy is but a small part of energy policy, it developed in interaction with the gradual setting of the internal electricity market. It also gave rise to a considerable body of expertise and research on renewable energy support schemes, their best designs and their effects. It is the interplay across these four processes—the promotion of electricity from renewable energy sources in Member States, the deployment of the internal electricity market as the backbone of EU energy policy, the simultaneous elaboration by the Commission of a EU perspective on renewable energy policy, and the development of expertise specialised on renewable energy policy—that we seek to explore in this book, as it played out in the evolution of FITs and of debates around them (Table 1.1).

Approaching Feed-in Tariffs as Socio-Technical Agencements

How to Retrace a European History of FITs?

FITs, as other RES-E support instruments, have received significant academic attention, either from the perspective of informing instrument choice and design (for instance by drawing lessons from empirical cases or

Table 1.1 Timeline and key dates

Key events in Member States		Key events at the EU level
Tariffs for RES-E based on avoided costs introduced in Germany and Denmark, on the basis of agreements between utilities and RES-E producers	1979	
	1988	European Council Recommendation on developing the exploitation of renewable energy sources
Stromeinspeisungsgesetz (Feed-in Law) in Germany	1991	Launch of Extern-E (External costs of Energy) European Research Network
Legislation on FITs in Denmark	1992	
FITs introduced in Spain	1994	
	1996	Directive 96/92/CE on the internal electricity market Green Paper on Renewable energy sources
	1997	White Paper on Renewable energy sources
TGCs introduced in Denmark	1999	
Erneuerbare-Energien Gesetz (Renewable Energy Law) in Germany	2000	
FITs introduced in France		
	2001	Directive 2001/77/CE on the promotion of electricity produced from renewable energy sources in the internal electricity market European Court of Justice states that FITs are not state aided in the Preusen Elektra v. Schleswag case
Introduction of FITs in the Czech Republic	2002	
	2003	Diffusion of results from Extern-E (External costs of Energy) European Research Network
Reform of the EEG in Germany	2004	
Announcement that FIT will be introduced in 2010 in the UK	2008	Energy-Climate Package
Drastic cuts of FITs for PV in Spain		
Revision of the EEG in Germany	2009	Directive 2009/28/CE on the promotion of the use of energy from renewable sources Renewable energy progress report

(*continued*)

Table 1.1   (continued)

Key events in Member States		Key events at the EU level
FITs introduced in the UK Revision of EEG in Germany Moratorium on FITs for PV in France Moratorium on FITs in the Czech Republic	2010	
FIT cuts in the UK	2011	
Reform of the EEG in Germany	2014	European Commission's Guidelines on State-aid for environmental protection and energy
	2015	Energy Union Package

by evaluating instruments in terms of efficiency and effectiveness) (e.g. Ménanteau et al. 2003; Midttun and Koefoed 2003; Haas et al. 2004, 2011; Sandén and Azar 2005, Frondel et al. 2008, 2010, Schmalensee 2012), or with a view to understanding their politics, that is how they came to be implemented in different countries and which interests and compromises shaped them (e.g. Lauber and Mez 2004; Meyer 2004; Jacobsson and Lauber 2006; Evrard 2010). The literature also includes analyses at the EU level (Mitchell 2010; Jacobs 2012; Solorio and Bocquillon 2017).

Here, we incorporate this literature (or, at least, a significant part of it) into our analysis, as one of the building blocks of our study. One of our objectives is indeed to understand to what extent it contributed to shaping FITs as well as how it informed, and also staged, debates about their design, effects, and place within the EU renewable energy policy arsenal. Our objective is not only to disentangle the politics of FITs in the European Union and in European countries, but also to understand how both the European Commission and experts in renewable energy economics and policy made sense—each in their own way—of the evolution of FITs and thereby contributed to shaping it.

For this purpose, we collected and analysed two corpuses of documents pertaining to the promotion of RES-E in Europe and published between the late 1980s and the early 2010s (Appendix). The first of these corpuses comprises EU documents directly related to renewable energy, mostly originating from the European Commission and the European Council[1] and released between 1986 and 2015 (relevant directives, Council resolutions, and Communications from the Commissions). The second corpus focuses on expertise, and includes grey literature (such as expert reports) and academic papers about FITs and renewable energy policy published between 1996 and 2013.[2] These documents were collected using a snowball method, and do not mean to be an exhaustive representation of the renewable energy policy literature; they are nonetheless diverse enough to cover the main fields of expertise involved (economics, policy analysis, innovation studies). We also established a chronology of renewable energy policy milestones at the EU level and in selected European countries (chiefly Denmark, Germany, Spain, France), drawing on survey reports (Mitchell et al. 2011; Jäger-Waldau 2013; REN21 2013) and on academic accounts (Meyer 2004; Lauber and Mez 2004; Jacobsson and Lauber 2006; Dinica 2008; Evrard 2010; Lauber and Schenner 2011), as well as on a detailed empirical knowledge of recent French renewable energy policies. In addition to this written material, we used interviews originally carried out as part of a study of FITs for photovoltaics in France. In this book, we refer to interviews with an energy economist, with a senior official in a large electric utility, and with two civil servants who worked on photovoltaic policy in France. These did not constitute the backbone of the analysis presented here, but they provided valuable inputs on the history of European energy policy and expertise, as well as a more detailed perspective on the actual challenges of tariff design.

Upon this basis, we established parallel timelines to trace the evolution of the issue of RES-E development in national policies, European legislation and policy principles, grey literature, and academic research. We then read the documents in the corpus closely so as to unravel, on the one hand, the European Commission's discourse on RES-E policy and FITs, and, on the other hand, the types of expertise mobilised on the subject and the evolutions in topics of research interest (Table 1.2).

Before explaining what this approach enabled us to do, it is important to outline what questions it does *not* address. First, we are not trying to

Table 1.2 Documents analysed by period and category

	EU documents	Grey literature	Academic literature
1978–1995	5	Ø	4
1996–2000	8	Ø	10
2001–2008	10	3	46
2009–2015	11	7	26
Total	34	10	86

evaluate FITs or provide policy recommendations on instrument design—something that numerous authors have done, and still do, better than us. Second, the account we produce gives a panoramic view, meaning that it does not go in depth into the details of either policy processes or knowledge production. The objective was to extract trends and discourses, bring them together to pinpoint synergies or tensions, and see how they play out in the concrete case of FITs. As a result, we do not map out extensively how expertise was formed and how it travelled, but focus on the identification of key themes, questions, and approaches, as well as on explicit references (in writing) or relations (in funding or commissioning) to EU policy processes. Similarly, our purpose is not to analyse the internal workings of European institutions: we thus do not retrace in detail disagreements, discussions and negotiations across and within European institutions—which would no doubt enrich the analysis, but provide too much detail for the format of a short book. In particular, the differences between different General Directorates in the Commission are not taken into account, as we have focused on the "finished products" of EU policy makings that are Communications from the Commission, Council resolutions, and Directives. We try to capture principles, guidelines, and an overall discourse that have been negotiated so as to reflect a shared EU approach—or that are presented as such in EU documentation—but in doing so, our intention is certainly not to convey the picture of the European Commission, or European institutions together, as a homogeneous, coherent actor: we only consider what remains when they manage to speak with one voice.

FITs as Agencements

By incorporating both policy documents and academic texts in our analysis, we approach the conception and evolutions of FITs as the joint product of political negotiations and policy implementation, on the one hand, and of economic reasoning and expertise, on the other. This view reflects a specific articulation of markets, expertise, and policy-making that is part of the liberalisation agenda, as outlined by Reverdy (2014, p. 56): the liberalisation of the energy sectors entails a recourse to economic expertise as a source of information and legitimacy for market regulation, embodied by independent regulatory authorities. It also resonates with the way the performativity tradition in economic sociology conceives of market devices (Callon et al. 2007; MacKenzie et al. 2008). Indeed, crucial to this tradition is the commitment not to consider theories about the economy as separated from the actual engineering and unfolding of economic activities. In particular, the concept of "socio-technical market agencements" or the notion of "agencing markets" help operationalise our empirical take on FITs.

Largely influenced by Science and Technology Studies (STS) and actor-network theory (ANT), the performativity tradition has sought to take economic knowledge seriously in sociological studies of markets. It considers the production of economic knowledge and theories as an integral part of the shaping of markets (and, more broadly, of economic activities), but, crucially, not as a description of the economy: the correspondence of economic theories and economic reality is the often-precarious result of materially enacted constructions that involve economic theories as much as practices (MacKenzie et al. 2008; Çaliskan and Callon 2009, 2010). This has notably translated into an attention to the "market devices" (Callon et al. 2007) that enable market transactions and market regulation, and in particular to the way they "frame" economic calculations, agencies and encounters (Callon 1998; Muniesa and Callon 2007).

Another key aspect emphasised by such studies is the contingency of market devices and framings: the organisation of economic activities is continuously being adjusted and renegotiated, because elements that were not accounted for (or that were "framed out") may turn out to matter—for example, externalities such as pollution or innovation (Callon 1998, 2007). This has led Callon to view market design as a process of collective experimentation in which the problems arising from the operation of markets are gradually accounted for and addressed through the "joint and coordinated advancement of knowledge and theoretical models on markets, on the one hand, and of market materials and institutional devices, on the other" (Callon 2009, p. 537). Instead of the notion of "experiment", which implies a degree of control and overseeing, Holm and Nielsen (2007) used the metaphor of cooking to describe the making of a market for individual quotas in Norwegian fisheries. They talked of the market as a "stew", and wondered if its production could easily be "attributed to some sort of cook" following a recipe—namely, economic theories on quotas (Holm and Nielsen 2007, p. 174). Their account provides a nuanced picture of the performativity of economic theories. They do not find a "master-plan" and a "cook" overseeing the process, but a rather messy process in which agencies and market devices (in this case, fishing quotas) are co-constructed: "a number of different materials went into the stew and they combined in unexpected and volatile ways" (Holm and

Nielsen 2007, p. 189). It is thus difficult, they argue, to maintain a "strict division between agency and devices; between cook and recipe" (Holm and Nielsen 2007, p. 190).

The notions of socio-technical market agencement (Barry 2001; Callon 2013), or of market agencing (Cochoy et al. 2016) originates in the work of Foucault and Deleuze, and was introduced in market studies to elaborate upon the notion of "market devices" and to invest it with deeper conceptual implications. Put briefly, agencements are defined as sets of elements that are put together in view to perform a specific function, and have agency precisely as a result of this careful articulation of heterogeneous parts. The term emphasises the intertwinement of *agencing* (arranging, putting together) and *agency,* and it also stresses the work required to put such arrangements together, make sense of them, and keep them together (thereby echoing ANT's conception of agency as distributed and full of surprises). Looking at the agencements that enable specific market transactions and economic activities then implies a focus on the combinations of institutional arrangements, material objects, and discourses and theories that organise economic activities. Crucially, theories and discourses about an agencement are part of it, because they inform its design and contribute to accounting for its working, making sense of it, and adjusting it.

To provide a thick description of FITs for electricity from renewable energy sources and of the way they make a difference, it is then not enough to explain their logic and consider their impact in terms of installed renewable electricity generation capacity. Approaching FITs as agencements outlines the diversity of ways in which they can be effectively set up, and the range of elements that they require to be put in place, tracked, and regulated. FITs are made of a combination of institutional arrangements (e.g. administrative procedures to obtain them, institutions in charge of their attribution and management, purchase agreements, levy for compensation, objectives for the deployment of renewable energy technologies, procedures for negotiating design…), of economic theories, methods and metrics (e.g. experience curves for technologies, definitions of costs and benefits, definitions of effectiveness and of efficiency, methods for evaluating investment risk, models of the energy mix, methods for establishing tariff rates…), and of a definition of RES-E as a tradable product (e.g. definition of renewable energy sources, technologies for converting them into electricity, connection to the grid, evaluation of resources…). The mixture can vary widely, but it always frames the goods that FITs can apply to, the modalities to purchase and sell them, the organisation of transactions, and to an extent the frame of reference against which to evaluate and reform (Cointe 2014, 2017). Crucially, since FITs are policy instruments, the mixture is ultimately the product of political objectives, negotiations and compromises—which does not necessarily downplay the role of economic expertise and theories, as these can be used as resources "in political struggles to define the rules of markets, property rights, and even the constitution of the calculative agents that are its participants" and serve as a frame of reference for market politics (Breslau 2013, p. 831).

This conceptual background appears particularly suited to the study of markets that are designed from scratch or deliberately amended or re-engineered, such as carbon markets (Callon 2009; MacKenzie 2009), fishing quotas (Holm and Nielsen 2007), capacity markets (Breslau 2013), or the European internal energy market (Reverdy 2014). The promotion of RES-E does not imply the creation and design of new markets per se, but it does involve some tinkering with electricity markets.

Our objective here is to take the economic logic of FITs and that of the EU internal electricity market programme seriously, while not forgetting that both FITs and the liberalisation agenda are shaped by political principles, negotiations and compromises—as Reverdy (2014) has shown limpidly for the construction of the prices for electricity and gas in the aftermath of liberalisation. With respect to this intention, one of the main merits of the notion of agencement is that it enables us to study market devices, but without focusing solely on transactions and markets. When talking of agencements, we do not have to confine our interest to economic activities, or to try and delineate markets and politics. Instead, we look at the varieties of issues, references and entities that are at stake in assembling a device—such as FITs—that organises a specific range of activities—here, investment in renewable electricity generation and the deployment of renewable energy technologies. Market transactions are part of the picture, but so are the organisation of the electricity system and the definition of energy policy objectives and institutions, and there is no need to draw firm lines between them to understand the history of FITs. Following Laurent (2015, p. 153), we do not study agencements "for the sake of the mere description of market exchanges", but rather "as analytical lenses for the characterisation of political and economic ordering". Agencements allow us to address such big issues precisely because while they are small, situated and concrete devices, political and economic orderings are at stake in their design.

What Can a History of FITs in the European Union Teach Us?

We approach FITs as agencements whose history and evolution was shaped in interactions across national renewable energy policies, EU policy, and economic expertise. As such, they provide an entry point to analyse how EU and Member States' renewable energy policies have unfolded in relation to the figure of the European Single Market and to trace the inter-weaving of theoretical and practical concerns in the process. Our analysis thus follows two main threads.

First, retracing the history of FITs as agencements provides a strategy to account for debates about European renewable energy policy instruments at the interface between environmental objectives and the ambition of market integration. We thus seek to contribute to the investigation of the constitution of a European political and economic space as it plays out in practice: focusing on one concrete type of instrument, we unpack the articulation of the promotion of renewable energy, on the one hand, with EU principles and ambition and, on the other hand, with the production of expertise about renewable energy policies and about their economic and political effects.

This situated take on the European project is in continuation of Barry's early analysis of the Single Market programme and of the project of har-monisation. Barry analyses these "as a means for developing and regulat-ing a European economic and social space, and as a project which would establish Europe as a space which could be acted upon, both by its author-ities and its subjects" (Barry 1993, p. 316). Our approach is also in con-tinuation of Barry's proposition that harmonisation "also increases the extent to which [objects and persons] can be known about and acted upon in a *European* way" (p. 322).

More recently, Reverdy (2014) has produced an extensive analysis of the constitution of energy markets as a domain of European economic and political regulation. He has retraced the strategies developed by govern-mental and industrial actors to accommodate or circumvent it. Laurent (2015) and Doganova and Laurent (2016) have described the constitution of a European political and economic ordering to deal with environmental issues through the building of specific hybrid political and economic agencements such as the establishment of criteria for the sustainability of biofuels and the definition of "best available techniques" for limiting pollu-tion. They argue that such agencements are exemplar of "a European way of acting on and through markets". Their study shows how the connection between environmental policy and market making in Europe plays out in specific policies. Similarly, when studying FITs in relation to EU renewable energy policy, what draws our attention is not just the development of renewable energy policies. We are as much interested in the issue of their integration within a European political and economic ordering, and their making into an object for academic research and for monitoring at the EU level.

In that sense, our study is an addition to analyses of the articulations of economic processes to politics and to technological developments, and of the devices through which such articulations and ordering are constituted. Energy appears as a particularly interesting object for such explorations, perhaps because technical economic and political consideration are closely imbricated in the organisation and regulation of such large systems as those for the provision of energy, as a range of STS works illustrate (e.g. Hecht 1998; Karnøe 2014; Yon 2014; Pallesen 2015; Silvast 2017; Cointe 2017). We argue that such studies can contribute to a more refined understanding of the diversity of economic activities, and particularly of their political and technological components.

This leads us to our second analytical thread. Historically, STS-informed studies of the economy have mostly focused on market and financial transactions as well as on consumption. Recently, they have started to explore other facets of economic activities. Two perspectives stand out. On the one hand, there has been an interest in what have been called "concerned markets", that is markets that are meant to fulfil other objectives and values than economic ones (Geiger et al. 2014), as well as on the capacity of markets to raise new issues and concerns (Callon 2007; Overdevest 2011; Cointe 2015) or to be arranged as tools for government (Ansaloni et al. 2017). On the other hand, it is being pointed out that economic activities cannot be reduced to financial speculation and market transactions. These dynamics, however relevant and important to study, may not be enough to account for the organisation of the economy. Muniesa et al. (2017) have thus suggested a study of the practices of capitalisation, defined as the part of economic activities that is framed in terms of investment and capital, and that values things not in terms of exchange values, but in terms of future returns. Particularly interesting is Muniesa et al.'s proposition that market valuation is only one kind of economic valuation and accounting, and that other forms of valuation, notably in terms of expected future returns, play a crucial role in economic activities.

In line with this recent literature, this book seeks to unpack the diversity of ways in which so-called "market-based" instruments can embody and enact diverse conceptions of economic engagement and diverse political orderings (e.g. who is responsible and what is relevant for regulating what, for designing instruments, for setting prices…). To this end, we will follow evolutions in the relative importance of competition and investment in the frames of references used to evaluate renewable energy policy instruments, especially FITs. We will also look into evolutions in the references mobilised to set feed-in tariffs. Reverdy has shown the importance of politics, regulatory frames, and to an extent bricolage in the establishment of energy prices following the liberalisation of the energy sector (Reverdy 2014). Similarly, we are interested in the debates around and the changes in the definition of a fair price for renewable electricity, as it relates to conceptions of what can constitute a "market basis" for policy and to shifts in political priorities, for example from market integration to climate change mitigation.

NOTES

1. EU Parliamentary documents were not analysed in as much detail because our objective was to trace the official and general EU discourse on renewable energy policy and feed-in tariffs that was constituted as an outcome of EU processes, rather than to analyse the political processes in which it was discussed, negotiated, and shaped.
2. Empirical work was carried out as part of a PhD that was defended in 2014, which is why the bulk of the corpus dates back from before 2013.

REFERENCES

Ansaloni, Matthieu, Pascale Trompette, and Pierre-Paul Zalio. 2017. Le marché comme forme de régulation politique. *Revue Française de Sociologie* 58 (3): 359–374.

Barry, Andrew. 1993. The European Community and European government: Harmonization, mobility and space. *Economy and Society* 22 (3): 314–326.

———. 2001. *Political machines: Governing a technological society*. London: Athlone Press.

Breslau, Daniel. 2013. Designing a market-like entity: Economics in the politics of market formation. *Social Studies of Science* 43: 829–851. https://doi.org/10.1177/0306312713493962.

Çaliskan, Koray, and Michel Callon. 2009. Economization, part 1: Shifting attention from the economy towards processes of economization. *Economy and Society* 38: 369–398. https://doi.org/10.1080/03085140903020580.

————. 2010. Economization, part 2: A research program for the study of markets. *Economy and Society* 39: 1–32. https://doi.org/10.1080/03085140903424519.

Callon, Michel. 1998. *The laws of the market*. Oxford: Blackwell/Sociological Review.

————. 2007. An essay on the growing contribution of economic markets to the proliferation of the social. *Theory, Culture & Society* 24: 139–163. https://doi.org/10.1177/0263276407084701.

————. 2009. Civilizing markets: Carbon trading between in vitro and in vivo experiments. *Accounting, Organizations and Society* 34: 535–548. https://doi.org/10.1016/j.aos.2008.04.003.

————. 2013. Qu'est-ce qu'un agencement marchand? In *Sociologie des agencements marchands. Textes choisis*, ed. Michel Callon et al., 325–440. Paris: Presses des Mines.

Callon, Michel, Yuval Millo, and Fabian Muniesa. 2007. *Market devices*. Oxford: Blackwell Publishing.

Cochoy, Franck, Pascale Trompette, and Luis Araujo. 2016. From market agencements to market agencing: An introduction. *Consumption Markets & Culture* 19 (1): 3–16.

Cointe, Béatrice. 2014. *The emergence of photovoltaics in France in the light of feed-in tariffs. Exploring the markets and politics of a modular technology*. Ph.D. thesis, Ecole des Hautes Etudes en Sciences Sociales, CIRED, Paris.

————. 2015. From a promise to a problem: the political economy of solar photovoltaics in France. *Energy Research and Social Science* 8: 151–161. https://doi.org/10.1016/j.erss.2015.05.009.

————. 2017. Managing political market agencements: Solar photovoltaic policy in France. *Environmental Politics* 26 (3): 480–501. https://doi.org/10.1080/09644016.2016.1269527.

Commission of the European Communities. 1996. Energy for the future: Renewable sources of energy. Green Paper for a Community strategy. Communication from the Commission. COM(96) 576 final. Brussels, 20 November 1996.

————. 1997. Energy for the future: Renewable sources of energy. White Paper for a Community Strategy and Action Plan. COM(97) 599 final. Brussels, 26 November 1997.

————. 1999. Electricity from renewable energy sources and the internal electricity market. Commission working document. SEC(1999) 470 final, Brussels, 13 April 1999.

————. 2004. The share of renewable energy in the EU. Commission report in accordance with Article 3 of Directive 2001/77/EC, evaluation of the effect of legislative instruments and other Community policies on the development of the contribution of renewable energy sources in the EU and proposals for concrete action. COM(2004) 366 final. Brussels, 26 May 2004.

————. 2008. The support of electricity from renewable energy sources. Commission staff working document accompanying document to the Proposal for a directive of the European Parliament and of the Council on the promotion of the use of energy from renewable sources {COM(2008) 19 final}, SEC(2008) 57, Brussels, 23 January 2008.

Couture, Toby, and Yves Gagnon. 2010. An analysis of feed-in tariffs remuneration models: Implications for renewable energy investment. *Energy Policy* 38: 955–965. https://doi.org/10.1016/j.enpol.2009.10.047.

Dinica, Valentina. 2008. Initiating a sustained diffusion of wind power: The role of public-private partnerships in Spain. *Energy Policy* 36: 3562–3571. https://doi.org/10.1016/j.enpol.2008.06.008.

Doganova, Liliana, and Brice Laurent. 2016. Keeping things different: Coexistence within European markets for cleantech and biofuels. *Journal of Cultural Economy* 9 (2): 141–156.

Eising, Rainer. 2001. Policy learning in embedded negotiations: Explaining EU electricity liberalization. *International Organizations* 56 (1): 85–120.

European Commission. 2011. Renewable energy: Progressing towards the 2020 target. Communication from the Commission to the European Parliament and the Council. COM(2011) 31 final. Brussels, 31 January 2011.

————. 2013. Renewable energy progress report. Report from the Commission to the European Parliament, the Council, The European Economic and Social Committee and the Committee of the Regions. COM(2013) 175. Brussels, 27 March 2013.

————. 2014. Guidelines on State aid for environmental protection and energy 2014–2020. Communication from the Commission. 2014/C 200/01.

————. 2015. A framework strategy for a resilient energy union with a forward-looking climate change policy. Communication from the Commission to the European Parliament, the Council, the European Economic and Social Committee, the Committee of the Regions and the European Investment Bank. Energy Union Package. COM(2015) 80 final. Brussels, 25 February 2015.

European Council. 1988. Council recommendation of 9 June 1988 on developing the exploitation of renewable energy sources in the Community. *Official Journal of the European Communities* L 160: 46–48.

European Parliament and Council. 1996. Directive 96/92/EC of the European Parliament and of the Council of 19 December 1996 concerning common rules for the internal market in electricity. *Official Journal of the European Communities* L 027: 20–29.

————. 2001. Directive 2001/77/EC of the European Parliament and of the Council of 27 September 2001 on the promotion of electricity produced from renewable energy sources in the internal electricity market. *Official Journal of the European Communities* L 283: 33–40.

———. 2009. Directive 28/2009/EC of the European Parliament and of the Council of 23 April 2009 on the promotion of the use of energy from renewable sources and amending and subsequently repelling Directives 2001/77/EC and 2003/30/EC. *Official Journal of the European Union* L 140: 16–62.

Evrard, Aurélien. 2010. L'intégration des énergies renouvelables aux politiques publiques en Europe. Ph.D. thesis, Sciences Po Paris, Paris.

Frondel, Manuel, Nolan Ritter, and Christoph M. Schmidt. 2008. Germany's solar cell promotion: Dark clouds on the horizon. *Energy Policy*36: 4198–4204. https://doi.org/10.1016/j.enpol.2008.07.026.

Frondel, Manuel, Nolan Ritter, Christoph M. Schmidt, and Colin Vance. 2010. Economic impacts from the promotion of renewable energy technologies: The German experience. *Energy Policy* 38: 4048–4056. https://doi.org/10.1016/j.enpol. 2010.03.029.

Geiger, Susi, Debbie Harrison, Hans Kjellberg, and Alexandre Mallard. 2014. Being concerned about markets. In *Concerned markets: Economic ordering for multiple values*, ed. Susi Geiger, Debbie Harrison, Hans Kjellberg, and Alexandre Mallard. Cheltenham: Edward Elgar Publishing.

Haas, Reinhaard, Wolfgang Eichhammer, Claus Huber, Ole Langniss, Arturo Lorenzoni, Reinhard Madlener, Philippe Ménanteau, Poul Erik Morthorst, Alvaro Martins, Anna Oniszk, et al. 2004. How to promote renewable energy systems successfully and effectively. *Energy Policy* 32: 833–839. https://doi.org/10.1016/S0301-4215(02)00337-3.

Haas, Reinhaard, Gustav Resch, Christian Panzer, Sebastian Busch, Mario Ragwitz, and Anne Held. 2011. Efficiency and effectiveness of promotion systems for electricity generation from renewable energy sources—Lessons from EU countries. *Energy* 36: 2186–2193. https://doi.org/10.1016/j.rser.2010.11.015.

Hecht, Gabrielle. 1998. *The radiance of France. Nuclear power and national identity after World War II*. Cambridge MA: MIT Press.

Holm, Peter, and Kåre Nolde Nielsen. 2007. Framing fish, making markets: The construction of Individual Transferable Quotas (ITQs). *The Sociological Review* 55: 173–195. https://doi.org/10.1111/j.1467-954X.2007.00735.x.

Hoppmann, Joern, Joern Huenteler, and Bastien Girod. 2014. Compulsive policy-making—The evolution of the German feed-in tariff system for solar photovoltaic power. *Research Policy* 43 (8): 1422–1441. https://doi.org/10.1016/j.respol.2014.01.014.

Jabko, Nicolas. 2006. *Playing the market. A political strategy for uniting Europe, 1985–2005*. Ithaca, NY: Cornell University Press.

Jacobs, David. 2010. Fabulous feed-in tariffs. *Renewable Energy Focus*, July–August 2010: 28–30.

———. 2012. *Renewable energy policy convergence in the EU. The evolution of feed-in tariffs in Germany, Spain and France*. Farnham: Ashgate.

Jacobsson, Staffan, and Volkmar Lauber. 2006. The politics and policy of energy system transformation—Explaining the German diffusion of renewable energy

technology. *Energy Policy* 34: 256–276. https://doi.org/10.1016/j.
enpol.2004.08.029.
Jäger-Waldau, Arnulf. 2013. *PV status report 2013*. JRC Scientific and Technical
Reports. Ispra: European Commission Joint Research Centre.
Karnøe, Peter. 2014. Large scale wind power penetration in Denmark—Breaking up
and remixing politics, technologies and markets. *Revue de l'Energie* 611: 12–22.
Lauber, Volkmar, and Lutz Mez. 2004. Three decades of renewable electricity
policies in Germany. *Energy and Environment* 15: 599–623. https://doi.
org/10.1260/0958305042259792.
Lauber, Volkmar, and Elisa Schenner. 2011. The struggle over support schemes
for renewable electricity in the European Union: A discursive-institutionalist
analysis. *Environmental Politics* 19: 127–141. https://doi.org/10.1080/096
44016.2011.589578.
Laurent, Brice. 2015. The politics of European agencements: Constructing a market of sustainable biofuels. *Environmental Politics* 24: 138–155. https://doi.
org/10.1080/09644016.2014.927190.
Lesser, Jonathan A., and Xuejuan Su. 2008. Design of an economically efficient
feed-in structure for renewable energy development. *Energy Policy* 36: 981–990.
https://doi.org/10.1016/j.enpol.2007.11.007.
Loiter, Jeffrey M., and Vicki Norberg-Bohm. 1999. Technology policy and renewable energy: Public roles in the development of new energy technologies. *Energy
Policy* 27: 85–97. https://doi.org/10.1016/S0301-4215(99)00013-0.
MacKenzie, Donald. 2009. Making things the same: Gases, emission rights and
the politics of carbon markets. *Accounting, Organizations and Society* 34:
440–455. https://doi.org/10.1016/j.aos.2008.02.004.
MacKenzie, Donald, Fabian Muniesa, and Lucia Siu. 2008. *Do Economists make
markets? On the performativity of economics*. Princeton: Princeton University
Press.
Ménanteau, Philippe, Dominique Finon, and Marie-Laure Lamy. 2003. Prices ver-
sus quantities: Choosing policies for promoting the development of renewable
energy. *Energy Policy* 31: 799–812. https://doi.org/10.1016/S0301-4215(02)
00133-7.
Meyer, Niels I. 2004. Renewable energy policy in Denmark. *Energy for Sustainable
Development* 8: 25–35.
Midttun, Atle, and Anne Louise Koefoed. 2003. Greening of electricity in Europe:
Challenges and developments. *Energy Policy* 31: 677–687. https://doi.
org/10.1016/S0301-4215(02)00152-0.
Mitchell, Catherine. 2010. Examining European political paradigms. In *The politi-
cal economy of sustainable energy*. Basingstoke: Palgrave Macmillan.
Mitchell, Catherine, Janet Sawin, Govind R. Pokharel, Daniel Kammen, Zhongying
Wang, Solomne Fifita, Mark Jaccard, Ole Langniss, Hugo Lucas, Alain Nadaï,
et al. 2011. Policy, financing and implementation. In *IPCC Special Report on*

Renewable Energy Sources and Climate Change Mitigation, ed. Ottmar Edenhofer, Ramon PichsMadruga, Youba Sokona, Kristin Seyboth, Patrick Matschoss, Susanne Kadner, Timm Zwickel, Patrick Eickemeier, Gerrit Hansen, Steffen Schlömer, and Christoph von Stechow. Cambridge and New York: Cambridge University Press.

Muniesa, Fabian, and Michel Callon. 2007. Economic experiments and the construction of markets. In *Do economists make markets? On the performativity of economics*, ed. David MacKenzie, Fabian Muniesa, and Lucia Siu. Princeton, NJ: Princeton University Press.

Muniesa, Fabian, Liliana Doganova, Horacio Ortiz, Alvaro Pina-Stranger, Florence Paterson, Alaric Bourgoin, Vera Ehrenstein, Pierre-André Juven, David Pontille, Başac Saraç-Lesavre, and Guillaume Yon. 2017. *Capitalization: A cultural guide*. Paris: Presses des Mines.

Overdevest, Chistine. 2011. Towards a more pragmatic sociology of markets. *Theory and Society* 40: 533–552. https://doi.org/10.1007/s11186-011-9149-1.

Pallesen, Trine. 2015. Valuation struggles over pricing—Determining the worth of wind power. *Journal of Cultural Economy* 9 (6): 527–540.

REN21 [Renewable Energy Policy Network for the 21st Century]. 2013. *Renewable 2013*. Global Status Report. Paris: REN21.

Reverdy, Thomas. 2014. *La construction politique du prix de l'énergie*. Paris: Presses de Sciences Po.

Sandén, Björn A., and Christian Azar. 2005. Near-term technology policies or long-term climate targets? Economy-wide versus technology-specific approaches. *Energy Policy* 33: 1557–1576. https://doi.org/10.1016/j.enpol.2004.01.012.

Schmalensee, Richard. 2012. Evaluating policies to increase electricity generation from renewable energy. *Review of Environmental Economics and Policy* 6: 45–64. https://doi.org/10.1093/reep/rer020.

Silvast, Antti. 2017. Energy, economics and performativity: Reviewing theoretical advances in social studies of markets and energy. *Energy Research & Social Science* 34: 4–12.

Solorio, Israel, and Pierre Bocquillon. 2017. EU renewable energy policy: A brief overview of its history and evolution. In *A guide to EU renewable energy policy: Comparing Europeanization and domestic policy change in EU member states*, ed. Israel Solorio and Helge Jörgens. Cheltenham: Edward Elgar Publishing.

Yon, Guillaume. 2014. L'économicité d'EDF. La politique tarifaire d'Electricité des France et la reconstruction de l'économie nationale, de la nationalisation au milieu des années 1960. *Politix* 105: 91–115. https://doi.org/10.3917/pox.105.0091.

CHAPTER 2

FITs and European Renewable Energy Policy Before 1996: A Tale of Two Beginnings

Abstract This chapter focuses on the period between the late 1970s and the adoption of the Directive on the internal electricity market in 1996. This period saw the appearance of feed-in tariffs as well as the emergence of a European Union approach to renewable energy policy, though without direct connections between the two. Cointe and Nadaï retrace the origins of FITs, starting from their introduction in Germany and Denmark as a tool to integrate wind power into existing electricity systems and showing how they turned into renewable energy support instruments. They then analyse early European renewable energy policy, stressing that it was guided by the ideal of a liberalised market as a device for economic and social optimisation.

Keywords Feed-in tariffs • European Union • Renewable energy policy • Internal electricity market • External costs

The first mechanisms akin to feed-in tariffs (FITs) appeared in the late 1970s. The first countries to introduce them in their legislation, Germany, Denmark, and Spain, did so in the early 1990s. During the 1980s and 1990s, the European Union (EU) started to engage in energy policy and, as part of this move, to consider renewable energy. This chapter is interested in the early stages of both FITs and European renewable energy policy. It focuses on a period stretching from the late 1970s to 1996, when

© The Author(s) 2018
B. Cointe, A. Nadaï, *Feed-in tariffs in the European Union*, Progressive Energy Policy, https://doi.org/10.1007/978-3-319-76321-7_2

the Directive on the internal electricity market was adopted, asserting the EU's role in ordering European electricity systems. To investigate these two distinct but simultaneous processes, we start from two questions:

- Where do FITs come from?
- Upon which bases was the EU's approach to renewable energy policy built?

FITs originate in mechanisms introduced in the late 1970s to accommodate increasing amounts of small-scale electricity. In Europe, such mechanisms were first set up on a voluntary basis in Germany and Denmark, which pioneered the development of wind power. The first section of this chapter retraces how these mechanisms initially organised the inclusion of electricity from renewable sources into the existing ordering of electricity systems by making it equivalent to electricity from other sources; it then follows their gradual incorporation into law as part of a renewable energy promotion agenda, and the associated evolution in the rationale for FITs.

During the 1980s and early 1990s, European institutions also started developing an agenda for renewable energy policy, as part of a broader European energy strategy that was built around the project of liberalised internal markets for energy. The second part of the chapter looks into the way European institutions, and the European Commission in particular, approached the promotion of electricity from renewable energy sources (RES-E) and developed a rationale for policy intervention with renewable energy based on the principles of the internal electricity market. The elaboration of this conception of RES-E policy, which drew on economic expertise, is crucial to understand the later evolutions of European renewable energy policy and of the European Commission's position on FITs.

This chapter thus seeks to understand how electricity from renewable energy sources was incorporated into the realm of electricity economics and politics. In particular, it attempts to delineate the frame of reference for the promotion of RES-E that emerged during this period. Early occurrences of FITs and the elaboration of a European discourse on RES-E development both contributed to the emergence of this frame.

The Origins of Feed-in Tariffs

By the late 1970s, the development of small-scale electricity generation—chiefly from wind power—created a need for integrating new forms of power generation into existing electricity systems. This drove the creation of mechanisms for feeding renewable electricity into the grid and remunerating it, which is how "proto-FITs" first appeared. Initially, these mechanisms were designed to allow small-scale electricity production and RES-E to blend into existing markets: they were agencements that served as "adapters" to include a new type of electricity product to be exchanged. This is clear from the reference used to set prices for this new kind of electricity. Tariffs were indeed first defined in relation to the avoided costs of fossil fuels, that is, the costs that would have been paid to generate the same amount of electricity from fossil fuels, in an attempt to make RES-E equivalent to conventional electricity. In later arrangements, they were calculated on the basis of average sales price for electricity. Gradually, these "proto-FITs" agencements became entangled in ambitions to promote renewable energy, and turned into agencements to this end.

The first "FIT-like" mechanisms originate in the US, where they appeared in the late 1970s in the form of avoided costs payment schemes. The 1978 Public Utility Regulatory Policies Act (PURPA) required that electric utilities purchase electricity produced by so-called qualifying facilities (i.e. specific small-scale electricity producers) at prices reflecting their long-term avoided costs (Lesser and Su 2008, p. 982; Loiter and Norberg-Bohm 1999). In this arrangement, state Public Utility Commissions were in charge of calculating avoided costs, and they used various methodologies for that (Loiter and Norberg-Bohm 1999). Some relied on forecast models of future fossil fuel and electricity prices, which led to over-estimations of avoided costs (Lesser and Su 2008).

In Europe, similar mechanisms based on the principle of avoided costs appeared around the same time. The main ones were introduced in Denmark and Germany to deal with the development of wind power production and the regulation of its integration to the grid.[1] In both of these countries, they started as regularly renegotiated agreements between electric utilities and wind power producers. Legislation incorporating FIT-like mechanisms into government regulation did not appear until the early 1990s.

Feed-in Tariffs in Denmark

Denmark introduced investment subsidies for wind turbines in 1979. By then, wind power had been mostly developed by neighbourhood cooperatives, and "utilities had little experience in handling dispersed, small-scale electricity systems such as wind turbines" (Meyer 2004, p. 28). To accommodate the resulting development of wind power, the

Association of Danish Electric Utilities and the Danish Wind Power association together with Danish wind turbine producers signed a first agreement in 1979, under governmental supervision (Meyer 2004; Evrard 2010). Utilities thereby agreed to purchase surplus electricity from wind power producers and to feed it to the grid at a price lower than that paid by consumers and determined on the basis of avoided fuel costs (Evrard 2010, p. 212).

Pressure from wind power producers led to successive renegotiations of the agreement throughout the 1980s. In 1984, a new voluntary agreement stated that utilities were to pay for 35% of grid connection costs for wind power, and to purchase wind power surplus at a rate of 85% of retail price (Evrard 2010, p. 213). In 1992, the government eventually introduced legislation regarding the grid connection of wind power and establishing feed-in tariffs, following the same mark-up model: the tariff was fixed at 85% of the utility production and distribution costs. The addition of a tax refund allowed for returns on investment of 10 to 15% for wind power projects, which was high enough to yield growth in installed capacity (Meyer 2004, p. 28).

Feed-in Tariffs in Germany

The emergence of support for RES-E followed a similar path in Germany. In the 1970s, German wind power producers had difficulty selling their surplus electricity. This led to the signature of an association agreement between the electric industry associations *Verband der Elektrizitätswirtschaft* (VDEW), *Verband der Industriellen Energie und Kraftwirstschaft* (VIK) and *Bundesverband der Deutschen Industrie* (BDI) in 1979 (Lauber and Mez 2004, p. 600). Like its Danish counterpart, the agreement required electric utilities to purchase RES-E according to the principle of avoided costs. However, utilities interpreted this principle in a restrictive way, leading to a much less generous framework than that provided by the PURPA in the US (Evrard 2010, pp. 246–247; Lauber and Mez 2004, p. 600).

At the time, the German Minister for the Economy was reluctant to encourage the formation of markets for renewable energy, judging them to rely on non-mature technologies (Evrard 2010, p. 247). However, by the late 1980s, there was a growing consensus in Parliament "that it was time to create markets for renewable energy technologies"

(Lauber and Mez 2004, p. 601); support programmes such as the 100/250 MW wind programme and the 1000 solar roofs programme were created, and proposals to establish feed-in tariffs for RES-E were formulated (Lauber and Mez 2004).

The *Stromeinspeisungsgesetz* (StrEG), or Feed-in Law, was eventually adopted in 1990. Moving away from the principle of avoided costs,[2] it set feed-in rates meant to reflect the external costs of conventional power generation (i.e. from fossil fuels and nuclear) in order to level the playing field between different sources of electricity (Lauber and Mez 2004; Lipp 2007, p. 5488). Utilities were required to connect renewable energy generators to the grid and to purchase it at 65%–90% of the price paid by consumers, in the same logic as the Danish system. This was a change in the logic of the *agencement*: previously designed merely to accommodate RES-E that substituted a small share of conventional power, it was now designed to include RES-E on an equal footing, recognising defects in the extant electricity system that needed correction. Then, feed-in mechanisms became an active part of renewable energy support policies.

In both Denmark and Germany, feed-in tariffs proved effective in promoting renewable electricity, and especially in promoting wind power. As a direct effect of the StrEG, installed wind power capacity in Germany grew from 20 MW in 1989 to over 1100 MW in 1995 (Lauber and Mez 2004, p. 602). This, however, did not happen without raising opposition from conventional electricity generators. Utilities turned to judiciary action at several levels, from Länder to the European Community, in attempts to roll back the law. In particular, in 1996, VDEW filed a complaint with the European Commission's Directorate General in charge of competition (DG Competition), invoking a violation of State aid rules (Lauber and Mez 2004, p. 603).

Early European Approaches to Renewable Energy Policy and Feed-in Tariffs

The institution of the first FIT schemes was not directly related to policies at the European level. All the same, the emergence of feed-in mechanisms took place at a time when European institutions were starting to consider renewable energy. Documents from the late 1980s attest to the European Community's burgeoning interest for renewables. In 1986, the European Council published a Resolution "concerning new Community energy policy objectives for 1995 and convergence of the policies of the Member States" (European Council 1986). Following this, it issued a Recommendation on "developing the exploitation of renewable energy sources" on 9 June 1988 (European Council 1988).

In the latter, the Council confirmed its 1986 "objective of continuing the development of new and renewable energy sources and of increasing their contribution to the total energy balance". It then stressed that the "development of renewable energy source requires appropriate legislative, administrative and financial measures", and recommended Member States "to introduce, where appropriate and necessary, legislation and/or administrative procedures which would help to overcome, on a non-discriminatory basis, obstacles to the exploitation of renewable energy sources" (European Council 1988). The Council further confirmed its consideration of the issue of the development of renewable energy in its Resolution of 13 December 1993 concerning the promotion of renewable energy in the Community.

European institutions also kept track of renewable energy policies in Member States, and in particular of their compatibility with the rules of the Common Market. The StrEG was thus notified to the European Commission for approval under State aid provisions. The Commission decided not to raise objections "because of its insignificant effects and because it was in line with the policy objectives of the Community" (Lauber and Mez 2004, p. 602), but planned to reconsider the matter two years later. By the time VDEW lodged its 1996 complaint, the Commission started to worry that feed-in tariffs might provide "excessive" minimum prices for wind power, given technological evolutions since 1990 (Lauber and Mez 2004). In a letter to the German Government following VDEW's complaint, the Commission thus "expressed doubts about the continued compatibility of the *Stromeinspeinsungsgesetz* with the Community State aid rules" (European Court of Justice 2000, §19).

The greatest source of concern was the calculation of the minimum purchase price for electricity generated from wind (European Court of Justice 2000, §19). The Commission was particularly worried that a minimum purchase price markedly higher than the cost of wind power generation might distort competition between Member States, and thereby prove incompatible with the Common Market. It proposed alternative methods for fixing the level of the minimum purchase price that would restore the compatibility of the StrEG with State aid rules:

"reducing the minimum purchase price for wind electricity, limiting the support mechanisms in time and/or according to electricity production, or calculating the purchase price on the basis of avoided costs" (European Court of Justice 2000, §21). The German government did not follow these suggestions: in Germany, FITs had evolved from agencements making RES-E equivalent to conventional electricity to agencements for the promotion of RES-E as a distinct type of electricity. The development of renewable energy had become an objective by itself, and paying a higher tariff for it was justified insofar as extant market prices did not take into account the different qualities of RES-E and conventional electricity, as reflected by the reference to external costs that became a keystone of the conception of renewable energy support in the 1990s.

AGENCING RENEWABLE ENERGY POLICY WITHIN THE CONSTRUCTION OF THE EUROPEAN INTERNAL ELECTRICITY MARKET

The shift to external costs as the benchmark for calibrating support to RES-E is crucial to understand the logic of European renewable energy policy in the 1990s and its relationship to the Common Market agenda. When the EU started elaborating principles for an EU-wide framework for renewable energy, the achievement of an internal market for electricity was at the top of its agenda. An EU renewable energy policy was thus supposed to contribute to market integration in the energy sector. This translated into a conception of support to renewable energy that articulated the principles of the Common Market, innovation, and environmental protection through a reference to the external costs of electricity production, which deserve elaboration here. In this conception, renewable energy policies were meant to contribute to the internalisation of external costs related to environmental protection and technological innovation; they were justified to the extent that they could contribute to correcting market failures without creating additional distortions. When defining its renewable energy strategy, the EU thus sought to determine common rules to ensure that RES-E support policies remained compatible with the principles and expected operation of a harmonised, integrated internal electricity market that still was only a project. This conception is explicitly stated in European institution documents, and it can also be traced in the economics literature.

Helping Renewables "Find a Place in the Market"

The rules of the internal market in electricity were set out in Directive 96/92/EC concerning common rules for the internal market in electricity (European Parliament and Council 1996), adopted in December 1996. The elaboration of the 1996 Directive fuelled discussions and debates about European energy policy, and the promotion of renewable energy was among the topics considered. In the first half of the 1990s, bases for the European approach to renewable energy deployment were sketched in several documents that closely articulated it with the principles of the internal electricity market.

The 1993 White Paper on "Growth, Competitiveness and Employment—the challenges and ways forward into the twenty-first century" presented "clean technologies" as key for future economic prosperity, and articulated competitiveness and environmental protection as complementary (Commission of the European Communities 1993). In 1995, a Green and a White Paper on Energy Policy (Commission of the European Communities 1995a, b) clearly outlined the logic along which the European Commission conceived support to renewable energy: environmental protection, technological development and competitiveness could go hand-in-hand, and they should be articulated through the operation of a common, liberalised European energy market. Developing a functioning internal market was the core focus of the European Community, and, it follows, a European response to energy policy challenges should revolve around this goal. As the 1995 White Paper stated, "market integration is the central, determining factor in the Community's energy policy" (Commission of the European Communities 1995b, p. 3).

To work "with the market rather than against it", the Commission advocated the "internalisation of external costs and benefits", especially when it came to environmental concerns, but also technological innovation (Commission of the European Communities 1995b, p. 31). The White Paper thus considered that "a significant proportion of new technological development will be driven by environmental considerations" (p. 13), and chiefly by the growth of renewable energy. The internalisation of externalities justified support programmes or subsidies to help renewable energy "find a place in the market" (p. 18) and technological developments turn into market products. A key requirement for such support mechanisms was to be as little harmful to competition as possible. Instruments for the promotion of renewable energy are thus conceived as corrective agencements that should contribute to the good functioning of the internal electricity market, and the non-competitiveness of renewable energy technologies is considered a signal of market distortions.

Renewable Energy Policy as a Correction of Market Failures

The notion that environmental protection and technological development ought to be achieved along with the operation of a liberalised market was very influential in the European Commission's approach to RES-E policy. Similar conceptions can be found in the literature about interactions across policy, environmental protection, and technological development published at the time (Jaffe and Stavins 1995; Wiser and Pickle 1998; Norberg-Bohm 1999). This literature viewed innovation in green technologies as a means to achieve better environmental protection; it argued that one way of fostering innovation in those sectors was to design incentives that compensated the lack of competitiveness of green technologies, thus enabling technological innovations to penetrate markets.

In such a perspective, policy intervention is acceptable insofar as it constitutes a provisional correction to so-called market failures. Support policies are supposed to expire once they have fulfilled their duty to re-establish the supposedly real, unbiased, level operation of the market. Since they are conceived as corrective devices, these policies themselves should generate as little additional distortions as possible: they are only meant to correct the imperfections that translate in the absence of RES-E. Incumbent energy sources are conceived as benefitting from artificial competitiveness as a result of the failure to take hidden costs and externalities into account and from institutional barriers and lock-ins (Unruh 2000, 2002). Policy intervention is then expected to establish a level playing field by reducing the difference between the costs of renewable and conventional energy, thereby reducing the differential in competitiveness. In short, "the basic premise of all renewable energy development policies is that they create demand that otherwise would not exist at desired levels *under current market conditions*" (Lesser and Su 2008, p. 983, emphasis added). This relies on policy-driven and policy-sustained markets agencements that economic theory sometimes struggles to qualify. For instance, some analysts consider them as "artificial market[s]", given that they are not based "on the voluntary decision of the consumers/voters" and "rely on a command and control approach of a planned economy" (Haas et al. 2011). Their "artificiality" as defined by economic theory stems from the fact that they rely on a demand and/or a supply that are not stabilised yet—for instance, Sandén and Azar (2005, p. 165) state that "government supported market formation has proved to be critical when a private demand for a product has yet to be formed [...] or when costs need to be brought down to be competitive on commercial markets".

Internalising Externalities to "Reveal the Ultimate Performance" of Renewable Energy Technologies

The correction of market conditions through public intervention finds justifications in economic theory, which may explain why it fitted so well in the European Commission's liberalisation agenda. It is considered as an internalisation of externalities related to environmental impacts and technological innovation; or, as Ménanteau et al. (2003, p. 55) put it, it may be "theoretically justified in two main ways: internalisation of environmental externalities and stimulation of technological change".

The first justification—internalisation of environmental externalities—implies "levelling the playing field" either by internalising the external costs of conventional electricity production using a carbon tax or setting a quota market like the EU-ETS, or by setting subsidies to reflect the positive environmental externalities of renewable energy. Different logics for the estimation of these externalities have prevailed over time. Initially, calculations were based on the specificities of fossil fuels, since renewable energy sources were supported at the "avoided cost of fossil fuel". As renewable energy policies developed, however, support was increasingly designed according to the specificities of renewable energy sources. Since the early 1990s, the Commission has also supported a Europe-wide research network seeking to define a stable methodology for the assessment of external costs and to carry out case studies from which to assess the value of external costs, the Extern-E project (Extern-E 2010).

The second theoretical justification for market correction—the need to stimulate technological change—implies that policy support should help spur decreases in the cost of renewable energy by accelerating innovation and learning processes until renewable energy generation technologies reach competitiveness. It is tied closely to the justification of support as internalisation of environmental externalities: supporting technological innovation in renewable energy technologies (RET) until they reach economic maturity is eventually supposed to level the playing field. The development of renewable energy technologies may indeed contribute to changing current knowledge, routine and practices in energy markets and institutions, as much as to making renewable energy technologies competitive with incumbent energy technologies. In this logic, innovation in renewable energy technologies needs to be stimulated and supported because the failure of the market to take into account externalities hinders the innovation processes that would bring renewable energy technologies to the market under supposedly unbiased conditions.

Support to the deployment of renewable energy technologies on the market is then expected to "stimulate a dynamic process that will reveal their ultimate performance" (Ménanteau et al. 2003, p. 57). In turn, this process is supposed to lead to the correction of market imperfections, in particular of environmental externalities due to the use of fossil fuel. In this conception, the ultimate objective of policy intervention is not to increase the share of RES-E in the energy mix per se, but to perfect market mechanisms so that they will eventually reduce negative environmental consequences and barriers to innovation by taking into account their full costs. Once externalities are internalised, RES-E should be valued according to its intrinsic performance—one key assumption thus being that it has such intrinsic performance.

CONCLUSION

Before 1996, renewable energy policy was still in its early stages, both in Member States and at the European level. All the same, the period stretching from the 1980s to the early 1990s was that of the emergence of agencements for the inclusion of RES-E within mainstream electricity grids, markets, and policy. Pioneering countries like Germany and Denmark instituted feed-in mechanisms that have since then drawn the contours of FITs as we know them. These first made wind power, and by extension RES-E, equivalent to conventional electricity, and later made it possible to promote it because of its distinct qualities. Meanwhile, the European Community worked towards the liberalisation of the energy sector, and began to consider renewable energy from this standpoint. It sketched the bases of a European approach to renewable energy policy. These bases consisted of general principles rather than of a detailed, concrete perspective on policy choices. They largely framed renewable energy policy as a tool for liberalisation and for the achievement of the internal electricity market: the promotion of RES-E was conceived in its capacity to correct market failures.

As far as we can trace, there were limited interactions between the emergence of FITs in Germany and Denmark and the first stages of European renewable energy policy principles. The main ones were probably the notification of the StrEG to the European Commissions, and a few years later, the expression of concerns from the Commission that FITs might constitute state aid. However, the centrality of the notion of "external costs" is a common trait of early FIT schemes and European rhetoric, and something that we also find in the academic literature. In Germany, it replaced "avoided costs of conventional electricity generation" as the benchmark to establish the level of FITs, thus allowing for the consideration of the characteristics of RES-E without being limited to the valuation practices of conventional power generation. In European Commission documents, the need to internalise externalities related to environmental impact and innovation was the main rationale for policy intervention to support RES-E, in line with justifications put forward by economic theory. This focus on externalities can be read as a mark of the influence of economic discourses and theories; but it can also be viewed as a rhetorical strategy to justify the promotion of RES-E in terms compatible with the framework of the internal electricity market, and to incorporate environmental and innovation objectives in this framework. At any rate, it remained a centrepiece of the European Commission's framing of RES-E policy and appreciation of FITs.

NOTES

1. "Avoided costs" based mechanisms were also used for co-generation for instance in Italy and France (Interview, energy economist 2013).
2. In 1989, the framework for electricity tariffs was modified so as to allow for compensation of RES-E generators above avoided costs (Lauber and Mez 2004, p. 601).

REFERENCES

Commission of the European Communities. 1993. *Growth, competitiveness, employment: The challenges and ways forward into the 21st century*. White Paper. COM(93) 700. Brussels, December.

———. 1995a. *For a European Union energy policy*. Green Paper. COM(94) 659 final. Brussels, 11 January 1995.

———. 1995b. *An energy policy for the European Union*. White Paper. COM(95) 682 final. Brussels, 13 December1995.

European Council. 1986. Council resolution of 16 September 1986 concerning new Community energy policy objectives for 1995 and convergence of the policies of the Member States. *Official Journal of the European Communities* C 241: 1–3.

——. 1988. Council recommendation of 9 June 1988 on developing the exploitation of renewable energy sources in the Community. *Official Journal of the European Communities* L 160: 46–48.

European Court of Justice. 2000. Opinion of advocate general Jacobs delivered on 26 October 2000, Case C-379/98 PreussenElektra v. Schleswag.

European Parliament and Council. 1996. Directive 96/92/EC of the European Parliament and of the Council of 19 December 1996 concerning common rules for the internal market in electricity. *Official Journal of the European Communities* L 027: 20–29.

Evrard, Aurélien. 2010. *L'intégration des énergies renouvelables aux politiques publiques en Europe.* Ph.D. thesis, Sciences Po Paris, Paris.

Extern-E. 2010. Extern-E Externalities of Energy—A Research Project of the European Commission. http://www.externe.info/externe_2006/exterpols. html. Accessed 12 Dec 2017, last modified 5 Oct 2010, consulted.

Haas, Reinhaard, Gustav Resch, Christian Panzer, Sebastian Busch, Mario Ragwitz, and Anne Held. 2011. Efficiency and effectiveness of promotion systems for electricity generation from renewable energy sources—Lessons from EU countries. *Energy* 36: 2186–2193. https://doi.org/10.1016/j.rser.2010.11.015.

Jaffe, Adam B., and Robert N. Stavins. 1995. Dynamic incentives of environmental regulations: The effects of alternative policy instruments on technology diffusion. *Journal of Environmental Economics and Management* 29: S43–S63. https://doi.org/10.1006/jeem.1995.1060.

Lauber, Volkmar, and Lutz Mez. 2004. Three decades of renewable electricity policies in Germany. *Energy and Environment* 15: 599–623. https://doi.org/10.1260/0958305042259792.

Lesser, Jonathan A., and Xuejuan Su. 2008. Design of an economically efficient feed-in structure for renewable energy development. *Energy Policy* 36: 981–990. https://doi.org/10.1016/j.enpol.2007.11.007.

Lipp, Judith. 2007. Lessons for effective renewable energy policies from Denmark, Germany and the United Kingdom. *Energy Policy* 35: 5481–5495. https://doi.org/10.1016/j.enpol.2007.05.015.

Loiter, Jeffrey M., and Vicki Norberg-Bohm. 1999. Technology policy and renewable energy: Public roles in the development of new energy technologies. *Energy Policy* 27: 85–97. https://doi.org/10.1016/S0301-4215(99)00013-0.

Ménanteau, Philippe, Dominique Finon, and Marie-Laure Lamy. 2003. Prices versus quantities: Choosing policies for promoting the development of renewable energy. *Energy Policy* 31: 799–812. https://doi.org/10.1016/S0301-4215(02)00133-7.

Meyer, Niels I. 2004. Renewable energy policy in Denmark. *Energy for Sustainable Development* 8: 25–35.

Norberg-Bohm, Vicki. 1999. Stimulating green technological innovation: An analysis of alternative policy mechanisms. *Policy Sciences* 32: 13–38. https://doi.org/10.1023/A:1004384913598.

Sandén, Björn A., and Christian Azar. 2005. Near-term technology policies or long-term climate targets? Economy-wide versus technology-specific approaches. *Energy Policy* 33: 1557–1576. https://doi.org/10.1016/j.enpol.2004.01.012.

Unruh, Gregory C. 2000. Understanding carbon lock-in. *Energy Policy* 28: 817–830. https://doi.org/10.1016/S0301-4215(00)00070-7.

———. 2002. Escaping carbon lock-in. *Energy Policy* 30: 317–325. https://doi.org/10.1016/S0301-4215(01)00098-2.

Wiser, Ryan H., and Steven J. Pickle. 1998. Financing investments in renewable energy: The impacts of policy design. *Renewable and Sustainable Energy Reviews* 2: 361–386. https://doi.org/10.1016/S1364-0321(98)00007-0.

CHAPTER 3

Tariffs, Quotas and the Ideal of Pan-European Harmonisation from 1996 to 2001

Abstract This chapter is interested in the period corresponding to the elaboration of the Directive on the promotion of renewable electricity, from 1996 to 2001. It looks at debates about renewable electricity policy at the European Union level as they became more specific: as national renewable electricity policies developed and diversified, attention started to focus on instrument choice. Having described the range of instruments in use, Cointe and Nadaï take the opposition between FITs and TGCs as a vantage point to analyse the Commission's vision and its discrepancies with national policies. They show how the Commission was guided by an ideal of harmonisation via the market that was not the priority in all Member States. They trace the emergence of similar debates in academic circles.

Keywords Feed-in tariffs • European Union • Tradable Green Certificates • Harmonisation

In 2001, a Directive "on the promotion of electricity from renewable energy sources and the internal electricity market" was adopted, defining a European frame for renewable energy policy. This chapter focuses on the period preceding the adoption of this directive: this was when the promotion of electricity from renewable energy sources (RES-E) really became a

© The Author(s) 2018   39
B. Cointe, A. Nadaï, *Feed-in tariffs in the European Union*, Progressive
Energy Policy, https://doi.org/10.1007/978-3-319-76321-7_3

matter of European attention and coordination. While Member States set up various strategies to encourage the development of RES-E, the European Commission elaborated its own conception of renewable energy policy, one that sought to reconcile policy intervention with the overarching objectives of market integration and harmonisation. Whereas the European Commission's position on renewable energy policy intervention had hitherto remained limited to rather general guiding principles, it now addressed the specifics of instrument choices.

Following a brief mapping of policy instruments for the promotion of RES-E available at the time and of the European Commission's renewable energy agenda, we focus on the debate opposing price-based support instruments such as feed-in tariffs (FITs) to quantity-based instruments such as Tradable Green Certificates (TGC). This opposition was structured in the discussions surrounding the elaboration of the 2001 Directive. The Commission clearly favoured TGCs, but this stance to an extent clashed with ongoing policy developments in Member States. We retrace the elaboration and consolidation of the European Commission's conception of what a good RES-E policy should be as it played out in this opposition between FITs and TGCs. This allows us to unpack the Commission's argumentation and the bases it relies upon. This is also a way to characterise both FITs and TGCs as distinct agencements, especially in the way they organise investment and competition and relate to the European single market framework.

Towards a European Strategy for Renewable Energy Policy

Following the adoption of the 1996 directive, the European Community reaffirmed its ambition to promote renewable energy and refined its approach to renewable energy policy, moving towards a more concrete perspective. The Commission released a Green and a White Paper on Renewable Energy Sources in 1996 and 1997, respectively (Commission of the European Communities 1996, 1997). The Green Paper drafted a political strategy articulated around four axes: a quantified target for increasing the share of renewable energy sources (RES) in energy consumption (namely its doubling by 2010), reinforced cooperation between Member States, reinforced European Community policies affecting RES development, and reinforced evaluation and monitoring of progress. It drew on previous European Community documents and legislation about renewable energy, on evaluations of the technical and economic potential of renewable energy and renewable energy technologies, and on a range of recently published energy scenarios.

Citing a forecasting study by the Commission entitled "European energy to 2020: a scenario approach", the 1996 Green Paper described four possible paths for the energy sector without specific renewable energy policies (Commission of the European Communities 1996). It referred to two studies exploring additional scenarios and taking into account renewable energy policy hypotheses: TERES and TERES II. The TERES studies confirmed the potential for renewable energy to contribute significantly to the European energy mix, provided that appropriate incentives were set. If renewable energy was to be significantly promoted in the European Community—and it was, if the Green Paper was any indication, then specific incentives were needed that would enable renewable energy sources to meet their potential.

Regarding these incentives, the Commission stayed in line with the market-centred approach of the internal electricity market framework. The introduction of renewable energy was considered to depend on the internalisation of external costs and on the correction of market failures then hindering the deployment of renewable energy technologies. Environmental protection and climate change mitigation were already crucial objectives, especially in the context of the then ongoing negotiation and adoption of the Kyoto Protocol, but the European Commission's discourse articulated them firmly to the good operation of the internal energy market. At this point, instrument choice did not appear to be the Commission's main concern, or at least the Commission did not express a clear preference for one option. According to the Green Paper, what mattered was not the type of instruments used, but their ability to provide sufficient transparency and stability to secure investments.

Various Options for Renewable Electricity Support Policies

The transposition of the 1996 Directive on the internal electricity market provided an opportunity for Member States to introduce or modify renewable energy policies. Objectives for the development of renewable energy had been established at the level of the European Union (EU), but Member States were free to design their own policy mix. As a result, renewable energy policy instruments varied widely from country to country. This calls for a brief overview of the forms that RES-E support can take, and an account of the specificity of generation-based policies such as FITs.

A relatively wide range of economic instruments can be used to help overcome obstacles to the development of RES-E: credit systems, taxes, state aids, standards and norms, support to R&D. Renewable energy support schemes can be quite sophisticated and include several types of incentives (Mitchell et al. 2011).

Renewable energy policy can, for instance, focus on renewable energy technologies improvement and cost reductions. To this end, two types of strategies can be combined: "technology push", which implies direct support to research and development in order to accelerate innovation, and "demand pull", which assumes that market deployment, for instance through subsidising investment in specific technologies, will trigger innovation and learning processes. Support can also directly target RES-E generation, and take the form of subsidies per kWh of renewable electricity produced. It is then expected to lead to increased installed capacity, thereby creating larger markets for renewable energy technologies.

Generation subsidies have gradually become the dominant form of regulatory support for renewable energy in Europe. They have rarely (if ever) been implemented on their own: in most countries, they are supplemented with diverse types of investment subsidies (renewable energy programmes, fiscal credits, local government subsidies...) and R&D support. In Germany and Denmark, the take-off of renewable energy was triggered by the combination of feed-in tariffs and investment subsidies or programmes, such as the German "1000 solar roofs programme". Yet, generation subsidies are the centrepiece in most renewable energy support schemes, EU guidelines, and academic research on the topic. This form of incentives is rather idiosyncratic to renewable energy policy. Indeed, RES-E generation subsidies can contribute to the different objectives that justify renewable energy market creation and development policies: they encourage investment in new capacity, they drive increases in RES-E production, innovation and cost reductions in RET, and, if designed to this end, they can internalise environmental and innovation external costs.

Focusing on the RES-E generation subsidies narrows down the range of instruments to four main categories: feed-in tariffs, feed-in premiums, renewable energy quotas, and tenders. The first two are "price-based" instruments—they set a level of remuneration for RES-E but provide no certainty as to the quantity of RES-E it will bring—while the two latter are "quantity-based"—they set a level of RES-E production to be achieved but provide no certainty as to the price that will be paid for to reach it.

Renewable energy quotas, called Tradable Green Certificates (TGCs) in Europe, set an amount of RES-E to be generated by allocating RES-E obligations to individual firms. Their logic is similar to that of carbon markets: each unit of renewable electricity generated gives right to a "tradable green certificate". The number of certificates available is determined by policy-makers, while their price is fixed on the market. To meet its obligations, a firm can produce RES-E, purchase it, or purchase certificates from firms who have produced more RES-E than they had to. RES-E production thus supposedly occurs where it is less costly. In tendering procedures, the government decides upon an overall objective, calls for tenders, and then selects the projects that will benefit from support on the basis of a predetermined range of criteria. Feed-in premiums work much like feed-in tariffs, except that they are connected to the fluctuation of prices on the electricity market. In such systems, RES-E producers receive a premium on top of the market price.

As of the end of the 1990s, this range of options was well represented throughout European countries. Germany and Denmark, as mentioned earlier, used feed-in tariffs (though Denmark was planning to replace them with tradable quotas by the end of the 1990s). Spain introduced feed-in premiums in 1997; the Netherlands and Sweden were pioneers in voluntary green pricing schemes in 1996, while France and the UK had opted for calls for tenders: the UK introduced the Non-Fossil Fuel Obligation (NFFO) in 1990, while France introduced the scheme "Eole 2005" in 1995. This diversity enabled the Commission to list and map policy options, even though it did not express preferences at this point.

The four categories of instruments differ in the ways they arrange competition: feed-in tariffs tend to create isolated and protected markets that are not exposed to competition,[1] while feed-in premiums connect RES-E prices to price fluctuations on the electricity market. Tenders and quotas also arrange rather distinct forms of competition: quotas imply that all types of RES-E compete on the same level; tenders set criteria along which competing investors will be ranked. Debates around renewable energy policy instruments have tended to focus on these four categories in both policy and academic circles. In the late 1990s, however, quotas and feed-in tariffs clearly dominated discussions, in particular in relations to the European Commission's concern for harmonising renewable energy policies across the EU—and, it follows, across the internal electricity market.

Orchestrating Renewable Energy Policy Across the European Union: The Commission's Harmonisation Agenda

Throughout the 1990s, renewable energy policies developed across EU Member States, but not in a particularly coordinated manner, as the diversity of available policy option suggests. In placing the promotion of renewable energy among its policy priorities, the main purpose of the European Commission was not to push Member States towards adopting renewable energy policies—most of them already had done so. Instead, the Commission sought to design a common framework able to ensure the coherence and stability of a wide variety of domestic policies, as well as their compatibility with an integrated European electricity market.

This purpose became increasingly explicit as the elaboration of European renewable energy policy progressed. As the share of renewable energy was expected to increase (political commitments had been taken to this end, notably via the Kyoto Protocol), so were the risks that a lack of coordination among Member States policies would distort trade and competition on the internal electricity market. The Commission's priority was to agree on a set of common rules for renewable energy policy, as expressed in its report on harmonization requirements for the internal electricity market:

> The contemporaneous existence of different support schemes appears likely to result in distortions of trade and competition. The role of renewables in the EU will clearly increase in the coming years, given the Kyoto commitments. Thus, potential market distortions will accordingly increase. Whilst the trade and competition distorting effects of different renewable support schemes is rather limited at present, given the limited EU market share of electricity from renewable sources, this negative effect appears likely to significantly increase in the coming years. In this light, it is appropriate to move towards the definition of some common rules in this area as rapidly as practicable. (Commission of the European Communities 1998, as quoted in Commission of the European Communities 1999, p. 5)

A 1999 Commission Working Document on "Electricity from renewable energy sources and the internal electricity market" (Commission of the European Communities 1999) provided the ground for discussions of these "common rules". To "[analyse] in detail the situation with regards to RES-E in the European Union", the Commission drew on a variety of sources:

> "Existing studies and reports on the design and functioning of current support mechanisms as well as on barriers other than financial, such as

administrative procedures and grid-system issues, were consulted. Furthermore, valuable information was received from Member States, on the basis of a questionnaire sent up by the Commission. Apart from the above investigations, discussions were held and/or comments received from many interested parties [...]." (Commission of the European Communities 1999, p. 45)

The resulting document provided an overview of the support schemes for RES-E generation deployed in EU Member States at the time. It distinguished between three types of incentives for RES-E generation: "fixed feed-in tariffs", "quota (competition-based) systems" and "fixed premium schemes", or feed-in premiums (FIP). This corresponds to the categories detailed above, except that tenders were not considered. The document then gave the Commission's assessment of their compatibility with EU rules and with the internal electricity market, before suggesting options for the EU-wide coordination and harmonisation of renewable energy policies.

The rationale for renewable energy policy as presented in this document remained the same as before: support was necessary to compensate for the temporary cost-disadvantages due to the non-internalisation of external costs of energy production; and it needed to be attractive enough to enable RES-E producers to enter the market, and stable enough not to deter investors. However, the objective of harmonisation as a way of ensuring the good operation of the internal electricity market clearly underlay the Commission's assessment of policy options. The Commission expressed a clear preference for competition-based quota systems that could be designed to contribute to the constitution of a unified and liberalised European electricity market. The working document is illustrative of a long-standing opposition between FITs and TGCs that first took shape during the elaboration of the 2001 Directive on the promotion of electricity from renewable energy sources and the internal electricity market, which the Commission intended as a step towards harmonisation.

The opposition between FITs and TGCs corresponds to a conception of the discussion of renewable energy policy as a matter of choosing between price-based (FITs or FIPs) and quantity-based (TGCs or tenders) instruments. This approach has been quite influential in policy discourses and in research. Since the results of tendering schemes adopted in the 1990s (the NFFO in the UK and Eole 2005 in France) had been deemed disappointing (Mitchell et al. 2011, p. 56),[2] the interest of governments and researchers focused on FITs and TGCs, which appeared as the two main options for organising renewable energy policy. The following section analyses the Commission's initial stance on the issue.

THE COMPATIBILITY OF FITS AND TGCS
WITH THE COMMISSION'S HARMONISATION AGENDA

In opposing FITs and TGCs, the European Commission was seeking to push for its own preferred solution, namely harmonisation of renewable energy policy based on a European TGC scheme, in a context where most Member States had opted for FITs. Upon examination, the Commission's position appears to be underpinned by a mixture of faith in competition as the best mechanism for solving problems and of reliance on economic concepts, logic, and rhetoric. FITs and TGCs are presented as two quite distinct ways of agencing economic activities. The following paragraphs attempt to disentangle this debate.

Feed-in Tariffs as Political Instruments Disconnected from Markets

By 2000, in addition to Germany, an increasing number of Member States had adopted, or were considering adopting, a form of FIT scheme: among them were Spain, France, Czech Republic, Greece, Portugal, and Luxembourg (Commission of the European Communities 2005). FIT schemes were already showing effects in the countries that had adopted them. The Commission recognised this, noting that "the highest levels of RES-generation increase have taken place in recent years in countries in which [a fixed feed-in tariff] operates" (Commission of the European Communities 1999, p. 16). In spite of this, it was clearly not in favour of FITs as a long-term instrument for supporting renewable energy. Instead, it advocated competition-based systems, a preference that can be directly linked to the belief in competition as the core organising mechanism for the European economic space that underlay the liberalisation agenda. The Commission's arguments against FITs all pointed to their disconnection from market competition. According to the Commission, this disconnection stemmed from the fact that FITs were prices that were *not* established in markets, and it had several implications.

By guaranteeing a fixed price for a determined period, FITs create a space for investments isolated from markets risks. They protect potential investors both from the fluctuations of electricity prices and from the evolutions of renewable energy technologies markets. Then,

> the actual price received by RES-E producers does not, necessarily, refer to any 'market price' for RES-E, nor necessarily takes account of falling RES-E production costs due to technological improvements. (Commission of the European Communities 1999, p. 12)

Indeed, translating this in the vocabulary of economic sociology, FITs are capitalisation devices (Muniesa et al. 2017) rather than market devices: they organise investments in RES-E generation as a source of future profits, not transactions of RES-E or competition among different types of electricity producers. Their level can be determined in reference to market-relevant valuation (reasoning in terms of external costs implies a reference to a market that is to be corrected), but it does not directly originate from the calculations of competing actors exchanging a product on a market. Rather, they tend to be determined in reference to investment costs, that is, to the profits that would be sufficient to entice investors to place money in RES-E generation.

The Commission considered this disconnection from market prices as a problem for three main reasons. First, it makes it difficult to articulate FIT-supported markets to others, which is a barrier to the constitution of an integrated, homogeneous European market space. Originally, feed-in tariffs were not designed to incorporate the possibility of trade with neighbouring countries, or to take into account evolutions in renewable energy technology markets (translating in cost reductions for RES-E), for instance.

Second, the level of FITs is set by public authorities. This implies that the incorporation of any new piece of information not previously accounted for in the tariff depends on government decisions. For the Commission, this was a major shortcoming. It limited the security provided by FITs, which "only exists so long as prices are not modified frequently" (Commission of the European Communities 1999, p. 15). It follows that, according to the Commission, FITs offered little flexibility and reactivity (since they would lose their main advantage if they were modified too often), leading to low levels of static efficiency,[3] and hence to a failure to "produce price reductions for RES-electricity" (Commission of the European Communities 1999, p. 16). The Commission considered regulatory authorities to be unable to react quickly enough to price reductions resulting from efficiency gains. Further, it viewed them as exposed to political difficulties and pressure from producers that would prevent them from adapting tariffs to prevent excessive profits. Third, a mechanism that did not depend on direct competition was also expected to provide little incentive for innovation, leading to low dynamic efficiency.

The Commission also stressed that the nature of feed-in tariffs generated uncertainty as to their compatibility with the EU state aid and internal market rules. For all of these reasons, while admitting that FITs could be an effective means to generate short-term increases in RES-E capacity (which they *de facto* were doing in several countries), the Commission did not envision them as a sustainable medium- to long-term option for RES-E support. When the European Court of Justice investigated the compatibility of the German *Stromeinspeisungsgesetz* feed-in tariff with European treaties for the PreussenElektra v. Schleswag case, the Commission stated its opinion that feed-in tariffs ought to be considered incompatible with the Treaty on the ground that they constituted State aid and could potentially hinder intra-Community trade (European Court of Justice 2001). FITs were perhaps agencements for organising the increase in RES-E generation capacity and its inclusion in the electricity system, but they were not agencements contributing to the constitution of an EU-wide liberalised electricity market.

Tradable Green Certificates as a Device for Harmonising Support to Renewable Electricity

By contrast, the European Commission advocated TGCs as agencements that were able to incorporate the promotion of RES-E into the constitution of the internal market for electricity, and to make it take part in this constitution. The hypothetical lower total costs of TGC were an argument, but they were not the only explanation for the Commission's preference for a quota-based system (Lauber and Schenner 2011). In fact, the Commission's preference had more to do with the fact that TGCs were perceived as more "market-compatible" [4] than FITs (Meyer 2003, p. 668), and market-compatible here means both favourable to competition and in line with the harmonisation objective underlying the internal electricity market project.

As opposed to FITs, quantity-based policies such as TGCs were expected to foster competition. Their proponents argued that they would direct subsidies towards least costly technological options and most favourable sites, and yield faster decreases in costs (Commission of the European Communities 1999), whereas FITs could encourage suboptimal investment:

> For some time, the RPS [Renewable Portfolio Standard, the equivalent of TGC in the US] was widely considered the least-cost approach to renewable energy development. The FIT was thought to be an expensive mechanism that did not stimulate sufficient competition within or across industry to bring costs down. (Lipp 2007, p. 5492)

By leaving price-setting to market mechanisms, they were also deemed less dependent on regulatory choice and less liable to regulatory capture than FITs.

Besides, TGCs appeared as more suitable devices in the hypothesis of a harmonization of RES-E support frameworks at the EU level, because TGC schemes appeared easier to link together than FIT schemes (Lauber and Schenner 2011; Commission of the European Communities 2004; Jacobsson et al. 2009). The similarities that TGCs shared with the Emission Trading Scheme (ETS) may also have played in their favour, especially considering that the EU, once a strong advocate of carbon taxation, had by then somewhat over-enthusiastically opted for a quota system for greenhouse gas emissions reduction (Branger et al. 2013).

TGCs best fitted with the Commission's focus on market integration and with its affirmed faith in the capacity of level market operation to bring about environmental protection and technological change. They were theoretically elegant solutions compatible with the ambition of achieving the internal electricity market because they had been designed to be so. FITs, on the other hand, were an effective but *de facto* solution with less firm theoretical grounding, and which the Commission may not have considered if they had not already been in place in several countries. The Commission was thus straightforward in its stance: it supported the progressive establishment of a harmonised community framework based on quantity-driven mechanisms (Lauber and Schenner 2011). More broadly, the Commission's purpose was to frame RES-E as a uniform good that would be tradable across a harmonised internal market without obstacles and differentiations. TGCs stemmed from the same logic as certificates of origins for RES-E, which would be implemented by the 2001 Directive to qualify "RES-E" as a category of electricity.

As Lauber and Schenner (2011) have shown, to support its choice, the Commission presented Tradable Green Certificates (TGC) systems as the most market-compatible solution, even though these had not been widely tested in Europe so far.[5] Indeed, most of the existing literature on TGCs was based either on theoretical analyses or on simulations of the expected impact of TGCs on RES-E development (Bergek and Jacobsson 2010).

As Lauber and Schenner noted,

> Given the strong belief in neoclassical economy, assumptions about the functionality of TGC were uncontested. Since no empirical experiences with TGC were available within Europe, theoretical assumptions could not be tested and went on faith. (Lauber and Schenner 2011, p. 517)

The Commission's preference for TGCs is indeed rooted in economic assumptions and in the language, logic, and concepts of economics. It seems to gain strength from its claimed grounding in economic theory, but how does it actually relate to academic discussions?

Prices and Quantities in Economic Theory

The opposition between FITs and TGCs frames the debates about renewable energy policy in terms of price-based versus quantity-based mechanisms—and this is the case in policy documents as well as in academic discussions. This distinction echoes the "price versus quantity" debates familiar in environmental regulation and environmental economics (Interview, senior official in electricity utility, 2012; Ménanteau et al. 2003). Since Weitzman's seminal 1974 paper on the subject, it is generally admitted that there is no absolute superiority of one type of policy over the other (Weitzman 1974, p. 478). As Ménanteau et al. write, transposing the debate to renewable energy policies,

> As with environmental policies, under the dual hypothesis of perfect information on the cost of renewable energy projects and zero transaction costs, price-based and quantity-based schemes produce very similar results. (Ménanteau et al. 2003, p. 62)

However, economic analysis can provide insights as to which of the two options is best suited to a specific regulation issue. Weitzman (1974) demonstrated that uncertainty was key while choosing between price and quantity instruments (Lecuyer and Quirion 2013). The main difficulty in terms of general welfare is to identify the most cost-efficient solution in situations of imperfect information. In the case of renewable energy development, in which neither the costs nor the benefits function are well-known, quantity-based instruments are often believed to allow for more control on total costs, because they set a cap on the amount of RES-E that will benefit from support:

> Controlling the cost of a renewable energy promotion policy based on a feed-in tariff is a classical difficulty associated with 'price-based' [...] rather than 'quantity-based' [...] environmental policies in situation of uncertainty over the costs and avoided damages function. (Finon 2008, p. 18, authors' translation)

Quantity-based instruments are also considered safer options because they are supposedly less subjected to the fluctuations of politics and to risks of regulatory capture. In general, economists would thus be inclined to consider quantity-based instruments as more cost-effective. As one interviewee explained, "in theory, if you are an economist, things are extremely simple; there are only two types of devices: quantity or price. If you know the quantity, you don't know the price, and if you know the price, you don't know the quantity". He continued, pointing out that economists would "always believe that an extremely competitive device based on quantities must logically lead to the least-cost result. Windfall profits seem unavoidable in a feed-in tariff system, since the regulator cannot be omniscient and know the tariff that correctly remunerates investment at every instant" (Interview, senior official in electricity utility, 2012).

This vision is schematic, and things were not as clear-cut: in fact, the matter was debated among economists (Hvelplund 2001; Interview, energy economist 2013). While infused with economic concepts and logics, the argument for TGCs was chiefly political, in that it reflected a priority given to harmonisation and to the achievement of the internal electricity market, as well as a firm belief in market competition as the best mechanism for the allocation of resources. The momentary pre-eminence of quantity-based instruments owed more to the European Commission's preferences, and to the perceptions of these preferences by Member States,[6] than to conclusions drawn from an economic expertise which was, in fact, not that homogeneous.

Harmonisation Postponed

The views of the European Commission were challenged even within European institutions, as Lauber and Schenner have explained in a discursive-institutionalist analysis of the debates over harmonisation of RES-E support (Lauber and Schenner 2011). Both the European Parliament and the Council opposed TGC-based harmonisation. The Parliament had previously argued for FIT-based harmonisation of RES-E support schemes (European Parliament 1996, 1998). Opposition from the Council related to subsidiarity: there were concerns that Members States would not be able to choose their own support schemes. In particular, Germany and Spain, both having FIT schemes running, were clearly against TGC-based harmonisation.

Lauber and Schenner show how the argumentation of the "anti-TGC discursive network" relied on two lines of argumentation. First, opposition to TGCs rejected the European Commission's economic frame of reference. Instead, it framed the problem as an environmental one, emphasising the observed effectiveness of FITs in driving RES-E deployment. This argumentation had little traction, because it did not address compatibility with market liberalisation, which was at the heart of the Commission's definition of the issue of renewable energy policy. The second line of argumentation against TGC-based harmonisation proved more effective, though it did not rely on an economic framing either. It invoked the subsidiarity principles and drew attention to the division of power between the EU and Member States: a pan-European TGC scheme would limit the ability of Member States to choose their own policy. As Lauber and Schenner put it, "many governments were simply determined to retain their own supports schemes" (Lauber and Schenner 2011, p. 519).

In the face of this opposition, the Commission eventually gave up, or rather postponed, its attempt at designing a harmonised community framework for renewable energy policy. The process launched by the 1996 Green paper resulted in Directive 2001/77/EC on the promotion of electricity from renewable sources and the internal electricity market, which set a EU target for the share of RES-E in electricity consumption, as well as indicative targets for Member States (European Parliament and Council 2001). In its article 4, the Directive explicitly stated that the objective of a harmonised framework had not been dropped, much to the contrary. RES-E policy was not mature enough yet to decide, but the issue was only postponed. The Commission kept its mandate to work towards this end:

> The Commission shall, not later than 27 October 2005, present a well-documented report on experience gained with the application and coexistence

of the different mechanisms referred to in paragraph 1. The report shall assess the success, including cost-effectiveness, of the support systems referred to in paragraph 1 in promoting the consumption of electricity produced from renewable energy sources in conformity with the national indicative targets referred to in Article 3(2). This report shall, if necessary, be accompanied by a proposal for a Community framework with regard to support schemes for electricity produced from renewable energy sources. (European Parliament and Council 2001, Article 4)

The perspective of a TGC-based harmonized EU framework had been a strong driver in the adoption of TGCs by Members States (especially Denmark, which changed from FITs to TGCs in order to align with European preferences). The fact that this perspective was now rejected into a distant future changed the picture. One of its consequences was effectively to disconnect the theoretical vision of a European internal electricity market from the more down-to-earth inclusion of RES-E in existing electricity markets and policy that was going on within Member States. Initially, the model of the internal electricity market and the future renewable electricity markets in Member States could be superimposed, since it was possible to envision that RES-E policy would develop in line with the principles of the integrated European market, and with the purpose to help integrate it. By 2001, the two appeared to have diverged: the creation of RES-E markets in Member States had emancipated from the principles stated at the European level—if indeed it had ever been subjected to them, and the integrated electricity market progressively turned from a blueprint into an ideal.

CONCLUSION

In the late 1990s, renewable electricity policy at the EU level became a matter of instrument choice. Earlier on, the European Commission's discourse on the promotion of RES-E had focused on the justification of policy intervention in the context of liberalisation of the internal electricity market. It had now become more specific, as the European Commission was pushing for a specific instrument design—namely, pan-European harmonised TGCs—as the device to ensure the compatibility of RES-E promotion with the internal electricity market.

In preparing the 2001 Directive, the European Commission refined its conception of renewable electricity policy, fleshing it up as a specific agencement of EU-wide, "market-based" support. This conception was framed by a neoclassical economics logic and rhetoric, in line with the principles that the Commission had put forward earlier in the 1990s: RES-E policy intervention was justified as an internalisation of externalities that would contribute to the achievement of the internal market for electricity. The proposal for a harmonised TGC schemes operationalised this conception and the associated emphasis on competition as the organising principle of European economic activity.

This chapter analysed the European Commission's position on RES-E policy, but it also pointed out how it clashed with the policies that Member States were actually setting up. The friction between European ambitions and national policies played out quite clearly in the opposition between FITs and TGCs in the preparation of the 2001 Directive: while TGCs appeared as the textbook solution favoured by the Commission, FITs were more widely adopted "on the ground". FITs also seemed to work quite well in the countries that had introduced them, at least in terms of increase in RES-E installed capacity. The FIT versus TGC debate opposed different justifications for RES-E policy: the Commission's market-based vision, in which RES-E policy was to be evaluated according to its contribution to the achievement of the internal market, and in which the focus was on correcting distortions to competition; and a vision focusing on the capacity of RES-E policy to address climate change, trigger reductions in technology costs, and encourage new industries, which was that of FIT supporters such as the renewable energy sectors, the German and Spanish governments, and the European Parliament. In addition, the debate also opposed different conceptions of EU actions in the field of energy. On the one hand, TGCs as promoted by the Commission corresponded to an ambition for harmonised EU action, coordinated by the European Commission and guided by the objective to achieve the internal market for electricity. On the other hand, FITs fell in line with a conception that stressed subsidiarity and the freedom of Member States to choose, design and steer their own RES-E policies. The latter conception prevailed in 2001, but the European Commission did not abandon its ambition for coordination and harmonisation.

NOTES

1. Though feed-in tariffs have been shown to foster competition among manufacturers, and in fact "shift competition from electricity price to equipment price" (Mitchell et al. 2011, p. 55).

2. The IPCC Special Report on Renewable Energy Sources notes that "while quotas and tendering systems theoretically make optimum use of market forces, government tendering system in particular have often had a stop & go nature that has not been conducive to stable investment conditions" (Mitchell et al. 2011, p. 56).
3. In the Commission Working Document, static efficiency is defined as the "ability to ensure that electricity is generated and sold at minimum costs" and dynamic efficiency as the "ability to foster innovation, thus, again, driving down costs" (European Commission 1999, p. 15).
4. Many analysts dismissed this notion that TGCs were any more "market-based" than FITs, recalling that both instruments relied on the creation of artificial and regulated market conditions (Hvelplund 2001; Meyer 2003; Haas et al. 2011). According to Meyer (2003, p. 668), "one (political) problem with the system is that a fixed price level does not conform to traditional market principles."
5. A tradable quota system had been implemented in the United States in the 1980s, where it was called "Renewable Portfolio System". Interestingly, it was designed to replace avoided-cost pricing that the Reagan era deregulation had made seem not market-based enough (Lauber 2004).
6. For instance, Denmark shifted from FITs to TGCs in anticipation of the advent of a harmonized TGC scheme like the one that the Commission pushed for.

References

Bergek, Anna, and Staffan Jacobsson. 2010. Are tradable green certificates a cost-efficient policy driving technical change or a rent-generating machine? Lessons from Sweden 2003–2008. *Energy Policy* 38: 1255–1271. https://doi.org/10.1016/j.enpol.2009.11.001.

Branger, Frédéric, Oskar Lecuyer and Philippe Quirion. 2013. The European Union Emission Trading System: Should we throw the flagship out with the bath water? *Working Paper 48*, Paris: CIRED.

Commission of the European Communities. 1996. *Energy for the future: Renewable sources of energy*. Green Paper for a Community strategy. Communication from the Commission. COM(96) 576 final. Brussels, 20 November 1996.

———. 1997. *Energy for the future: renewable sources of energy*. White Paper for a Community Strategy and Action Plan. COM(97) 599 final. Brussels, 26 November 1997.

———. 1998. Commission report to the Council and the European Parliament on Harmonization requirements. Directive 96/92 concerning rules for the internal market in electricity. COM(1998) 167, Brussels, 16 March 1998.

———. 1999. *Electricity from renewable energy sources and the internal electricity market*. Commission working document. SEC(1999) 470 final, Brussels, 13 April 1999.

————. 2004. *The share of renewable energy in the EU.* Commission report in accordance with Article 3 of Directive 2001/77/EC, evaluation of the effect of legislative instruments and other Community policies on the development of the contribution of renewable energy sources in the EU and proposals for concrete action. COM(2004) 366 final. Brussels, 26 May 2004.

————. 2005. *The support of electricity from renewable sources.* Communication from the European Commission. COM(2005) 627 final. Brussels, 07 December 2005.

European Court of Justice. 2001. *PreussenElektra v. Schleswag.* Judgement of the Court, 13 March 2001, Case C-379/98.

European Parliament. 1996. Mombaur own-initiative report on a community action plan for renewable energy sources. *Official Journal of the European Communities* C 211. http://eur-lex.europa.eu/legal-content/EN/TXT/PDF/?uri=OJ:C:1996:211:FULL&from=EN. Accessed 21 Dec 2017.

————. 1998. Report on network access for renewable energies—Creating a European directive on the feeding in of electricity from renewable sources of energy in the European Union; rapporteur Linkohr. A4-0199 and 98, PE224.949.fin, 26 May.

European Parliament and Council. 2001. Directive 2001/77/EC of the European Parliament and of the Council of 27 September 2001 on the promotion of electricity produced from renewable energy sources in the internal electricity market. *Official Journal of the European Communities* L 283: 33–40.

Finon, Dominique. 2008. L'inadéquation du mode de subvention du photovoltaïque à sa maturité technologique. *Working Paper 2008–09*, Paris: CIRED.

Haas, Reinhaard, Gustav Resch, Christian Panzer, Sebastian Busch, Mario Ragwitz, and Anne Held. 2011. Efficiency and effectiveness of promotion systems for electricity generation from renewable energy sources—Lessons from EU countries. *Energy* 36: 2186–2193. https://doi.org/10.1016/j.rser.2010.11.015.

Hvelplund, Frede. 2001. Political prices or political quantities? A comparison of renewable energy support systems. *New Energy* 5: 18–23.

Jacobsson, Staffan, Anne Bergek, Dominique Finon, Volkmar Lauber, Catherine Mitchell, David Toke, and Ariel Verbruggen. 2009. EU renewable energy support: Faith or facts? *Energy Policy* 37: 2143–2146. https://doi.org/10.1016/j.enpol.2009.02.043.

Lauber, Volkmar. 2004. REFIT and RPS: Options for a harmonised Community framework. *Energy Policy* 32: 1405–1414. https://doi.org/10.1016/S0301-4215(03)00108-3.

Lauber, Volkmar, and Elisa Schenner. 2011. The struggle over support schemes for renewable electricity in the European Union: A discursive-institutionalist analysis. *Environmental Politics* 19: 127–141. https://doi.org/10.1080/09644016.2011.589578.

Lecuyer, Oskar, and Philippe Quirion. 2013. Can uncertainty justify overlapping policy instruments to mitigate emissions? *Ecological Economics* 93: 177–191. https://doi.org/10.2139/ssrn.2192517.

Lipp, Judith. 2007. Lessons for effective renewable energy policies from Denmark, Germany and the United Kingdom. *Energy Policy* 35: 5481–5495. https://doi.org/10.1016/j.enpol.2007.05.015.

Ménanteau, Philippe, Dominique Finon, and Marie-Laure Lamy. 2003. Prices versus quantities: Choosing policies for promoting the development of renewable energy. *Energy Policy* 31: 799–812. https://doi.org/10.1016/S0301-4215(02)00133-7.

Meyer, Niels I. 2003. European schemes for promoting renewables in liberalised markets. *Energy Policy* 31: 665–676. https://doi.org/10.1016/S0301-4215(02)00151-9.

Mitchell, Catherine, Janet Sawin, Govind R. Pokharel, Daniel Kammen, Zhongying Wang, Solomne Fifita, Mark Jaccard, Ole Langniss, Hugo Lucas, Alain Nadai, et al. 2011. Policy, financing and implementation. In *IPCC special report on renewable energy sources and climate change mitigation*, ed. Ottmar Edenhofer, Ramon PichsMadruga, Youba Sokona, Kristin Seyboth, Patrick Matschoss, Susanne Kadner, Timm Zwickel, Patrick Eickemeier, Gerrit Hansen, Steffen Schlömer, and Christoph von Stechow. Cambridge and New York, NY: Cambridge University Press.

Muniesa, Fabian, Liliana Doganova, Horacio Ortiz, Alvaro Pina-Stranger, Florence Paterson, Alaric Bourgoin, Vera Ehrenstein, Pierre-André Juven, David Pontille, Başac Saraç-Lesavre, and Guillaume Yon. 2017. *Capitalization: A cultural guide*. Paris: Presses des Mines.

Weitzman, Martin L. 1974. Price vs. quantities. *The Review of Economic Studies* 41: 477–491.

CHAPTER 4

2001–2008: European-Scale Experimentation in Renewable Energy Policy-Making

Abstract The chapter analyses two evolutions in European renewable electricity policy between 2001 and 2008. The first was a sophistication of renewable electricity support schemes, in particular feed-in tariffs. The second was the consolidation of expertise about renewable electricity policy. Cointe and Nadaï describe how feed-in tariffs were re-arranged in their design and in their relations to European law. In parallel, they trace the expansion of literature on renewable electricity policy. This production of expertise was partly driven by the European Commission and its will to coordinate renewable electricity policy across Europe via regular assessments. The chapter ends on an analysis of the validation of feed-in tariffs as "market-based" by the Commission, arguing that it also reflected a shift in focus from competition to investment.

Keywords Feed-in tariffs • European Union • Erneuerbare-Energien Gesetz • Experimentation • Innovation trajectories

The 2001 Directive provided a legal framework for the deployment of electricity from renewable energy sources (RES-E) in the European Union (EU). It introduced three new features in European renewable energy policy. First, it defined an EU-wide objective of 22% of gross electricity consumption produced from renewable energy sources by 2010. Second,

© The Author(s) 2018   59
B. Cointe, A. Nadaï, *Feed-in tariffs in the European Union*, Progressive
Energy Policy, https://doi.org/10.1007/978-3-319-76321-7_4

it derived indicative targets for each Member State from this overall target; while non-binding, they provided a reference for assessing progress. Third, it gave the Commission the responsibility to review progress in renewable energy policy in Member States, and to publish reports to this end, the first one after five years.

The new Directive then drove the reinforcement and sophistication of RES-E support schemes in Member States. Some, such as Germany, Spain or France, opted for feed-in tariffs (FIT), while others like the UK, Sweden or Denmark set up quota-based systems. In the process, both types of instruments were adapted and started to diverge from the ideal types they derived from. The European Commission paid sustained attention to these developments, and this contributed to the structuration of renewable energy policy as a topic for academic research. A body of literature developed at the intersection of several streams of research, mainly in economics, to document and inform the evolution of RES-E support schemes. It paid particular attention to the respective virtues and flaws of FITs and Tradable Green Certificates (TGCs). Expertise about RES-E policy developed alongside a Europe-wide political process aiming at harmonising or at least coordinating domestic policies.

During this period, the conception of FITs evolved both in practice and in theory. FIT schemes turned out quite successful in driving increases in installed RES-E generation capacity, and they became increasingly considered as market-based instruments. A combination of theoretical and practical refinements led to their gradual validation in the literature and, eventually, by the European Commission. This chapter is precisely interested in this change of status of FITs and their gain of what we could call "market legitimacy" between the two renewable energy directives of 2001 and 2008.

A RE-AGENCEMENT OF FEED-IN TARIFFS

Contemporaneously with the adoption of the 2001 Directive, two policy and legal developments contributed to a redefinition of FITs, in particular regarding how their level related to market dynamics and to political decisions. The first of these developments was the adoption of the Renewable Energy Law in Germany in 2000, which introduced a much more sophisticated FIT scheme than what had hitherto existed. The second was the European Court of Justice's ruling on PreussenElektra versus Schleswag, which stated that the FITs introduced by the German *Stromeinspeisungsge-setz* (StrEG) in the 1990s did not constitute State aid.

The German Renewable Energy Law: Articulating FITs to Market Dynamics

Germany adopted the *Eneuerbare-Energien Gestez* (EEG) or Renewable Energy Law in 2000. This law reformed FITs in depth, turning them into much more sophisticated instruments than they used to be, and reconfiguring the articulation between political decisions and market information in their calibration.

The EEG drew inspiration from local FIT schemes for solar power that existed in some German regions to introduce three major modifications to FITs (Jacobsson and Lauber 2006, p. 267). First, tariffs were now guaranteed for twenty years, thus increasing investors' security. Second, the EEG also introduced technology-specific FIT rates, meaning that RES-E generated using different technologies would benefit from different levels of FITs. The purpose was to promote all types of renewable electricity sources, and not just wind power, which up to this point had been the main recipient of FITs. This was particularly relevant for solar power, which then remained much more expensive than many renewable energy sources: EEG FITs made "solar cells [...] an interesting investment option for the first time" (Jacobsson and Lauber 2006, p. 267). Last, the EEG introduced annual decline rates for FITs for new installations. They were intended to reflect the cost reductions triggered by the learning effects expected along market formation, and so they were also technology-specific. For instance, the FIT for photovoltaics was set to decline by about 5% every year.

Jacobsson and Lauber (2006) relate that the motivations of the Social Democrats in pushing the reform were industrial policy interests and the will to create employment opportunities (Jacobsson and Lauber 2006; Mitchell et al. 2011). The Social Democrats feared that "the 1998 liberalisation of the energy market would lead to a long-term decline in employment in the energy sector and in the associated capital goods industry" (Jacobsson and Lauber 2006, p. 267). However, in its wording, the EEG "repeated the Feed-in law's explicit commitments to take external costs into account" (Jacobsson and Lauber 2006, p. 267).

Perhaps not reflecting the politics behind the elaboration of the EEG, the explanatory memorandum attached to the EEG justified the new FIT using market-oriented arguments echoing the European Commission's justification for renewable energy policy: the internalisation of external costs via the polluter-pays principle, the need to level competition in a context where conventional energy sources remained heavily subsidised, and the will to "break the vicious circle of high unit costs and low production volumes" typical of renewable energy technologies (Jacobsson and Lauber 2006, p. 268; Federal Ministry for the Environment 2000). FITs were presented as instruments to improve electricity markets by taking environmental externalities into account, by re-establishing fair competition between different energy sources, and by lifting the barriers to investments hindering technological innovation and deployment. This justification in fact fits within the frame of "German Ordo-Liberalismus", which "stresses the need for the State to create a market framework for competition and to prevent monopolistic or oligopolistic power, and accordingly to prevent restrictions of market access, promote competition, and take measure against negative external effects" (Lauber 2004, p. 1406).

This conception can be traced in the way the EEG arranged support to RES-E. Within the EEG framework, FITs were not conceived as government-fixed prices to subsidise renewable energy production so much as dynamic devices to correct the market imperfections and barriers hindering the development of renewable energy. EEG FITs were designed to maximise investor's security—which was understood to guarantee the effectiveness of the instrument, but also to incorporate information on the costs of various renewable energy technologies, in a static (via technology differentiation) and a dynamic (via annual decline) manner.

In this manner, the new agencement of FITs set up in the EEG separated their calibration from politics. The level and evolution of support was tailored to the technical and market characteristics of each RES-E generation technology available and was made to depend primarily on these characteristics. Indeed, annual decline rates determined according to the innovation trajectories of various technologies were supposed to enable a "self-regulation" of FITs according to economic parameters. In theory, such FITs relied on market information about the costs of RES-E generation technologies, not on the arbitrariness of political adjustments: in that sense, they were "market-based" agencements supposedly protected from regulatory capture or political uncertainties. This contrasts sharply with the picture of FITs that the European Commission drew in its 1999 Working Document (see Chap. 3).

PreussenElektra v. Schleswag: Establishing the Compatibility of FITs with the Single Market

Until 2002, the European Commission's General Directorate for Competition expressed doubt regarding the compatibility of the EEG with EU rules (Jacobsson and Lauber 2006, p. 267). At the EU level, the legitimacy of FITs as market-based instruments, as opposed to State aid, was not fully established. The European Court's ruling on PreussenElektra v Schleswag in March 2001 was an important step in the legitimation of FITs (European Court of Justice 2001). It prevented the Commission from disqualifying FITs on the grounds of State aid regulation (Lauber and Schenner 2011).

The dispute was not about the FITs instituted by the EEG, but about the previous scheme introduced by the StrEG. The issue that had been raised by the German utility PreussenElektra was the compatibility of the purchase obligation of electricity from renewable energy sources created by the StrEG with the Treaty of the European Community. The Court was asked to state on two aspects: did the purchase obligation constitute State aid? And was it equivalent to a quantitative restriction on imports, and thus a distortion of trade between Member States incompatible with the principles of the Common Market (European Court of Justice 2000, 2001)?

The European Court of Justice rejected PreussenElektra's appeal to recognise the FIT introduced by StrEG as illegal state aid. It based its decision upon previous case law according to which State aid was to be interpreted in a restrictive manner as aid financed through State resource. While recognising that the StrEg purchase obligation was "capable, at least potentially, of hindering intra-Community trade", the Court stated that it was not incompatible with Treaty rules, on the grounds that it aimed at protecting the environment and that the particular features of the electricity market made it "difficult to determine [the origin of electricity] and in particular the source of energy from which it was produced" (European Court of Justice 2001, §17, 79). The European Court of Justice thereby sanctioned FITs as an acceptable means of support for RES-E development (Evrard 2010; Hvelplund 2001; Lauber 2004):

1. Statutory provisions of a Member State which, first, require private electricity supply undertakings to purchase electricity produced in their area of supply from renewable energy sources at minimum prices higher than the real

economic value of that type of electricity, and, second, distribute the financial burden resulting from that obligation between those electricity supply undertakings and upstream private electricity network operators do not constitute State aid within the meaning of Article 92(1) of the EC Treaty [...].

2. In the current state of Community law concerning the electricity market, such provisions are not incompatible with Article 30 of the EC Treaty. (European Court of Justice 2001)

Each in its own way, the European Court of Justice's ruling and the EEG both contributed to evolutions in the conception of FITs, and in particular to evolutions in their articulation to market processes and information—so in the way FITs organised the economic development of RES-E. These evolutions continued throughout the 2000s, in a joint process of intensification of expertise and research on renewable energy policy and of multiplication and sophistication of RES-E support instruments, especially FITs. This process can be likened to the constitution of a "network of experimentation" (Callon 2009), that is a network linking sites of research, business and policy-making and allowing experience, practices and proposals to circulate. It translated in theoretical and practical refinements that framed renewable energy support as an area of policy-making and as a research topic. We argue that it contributed to the Commission's eventual validation of FITs.

Experimenting with Renewable Energy Policy

The Constitution of a Field of Research and Expertise About Renewable Energy Policy

Abundant literature on RES-E policy was published in the 2000s. Most of the time, it was explicitly policy-relevant: research and expertise accompanied the development of renewable energy policy, and informed and/or influenced the Commission's monitoring and assessment of Member States policies.

A form of common perspective emerged from these interactions. Notions such as experience curves, learning-by-doing, or static and dynamic efficiencies became particularly influential in the design and conception of support instruments. The articulation between stimulating innovation and internalising environmental costs largely remained at the core of RES-E support, as suggested by an influential paper by Ménanteau et al. (2003):

Whatever the system chosen, the role of the public authorities is quite specific: to stimulate technical progress and speed up the technological learning processes so that ultimately renewable energy technologies will be able to compete with conventional technology, once the environmental costs have been internalised. (Ménanteau et al. 2003, p. 799, emphasis added)

The literature we have analysed here mainly explored the role of policy instruments in achieving the twin objectives of stimulating innovation and nurturing markets for renewable energy. We identify four main perspectives that have been convoked, and occasionally combined, to this end:

- a perspective largely influenced by *environmental economics* and literature on *environmental regulation*, with which it shares an interest in the internalisation of external costs and in public goods, as well as a concern for climate change mitigation: for example, Jaffe and Stavins 1995; Lecuyer and Quirion 2013; Ménanteau et al. 2003; Ringel 2006; Schmalensee 2012;
- a perspective focused on *innovation* and interested in the identification of drivers of technological change: for example, Freeman 1996; Foxon and Pearson 2008; Jamasb 2007; Midttun and Gautesen 2007; Papineau 2006; Schilling and Esmundo 2009;
- a perspective focused on *investment* and interested in the conditions for the reduction of investment risk and for market development: for example, Awerbush 2000; Couture and Gagnon 2010; Dinica 2006; Lüthi and Wüstenhagen 2012;
- an institutional and policy-making perspective, with a focus on *environmental policies* and on the institutional and political drivers of their adoption, design, and implementation: for example, Haas et al. 2004, 2008, 2011; Bergek and Jacobsson 2010.[1]

These research streams already had a history largely independent from their new object of study, though not necessarily independent from one another. They had addressed RES-E support more or less directly during the 1990s, but papers on the topic had remained relatively isolated (e.g. Freeman 1996; Wiser and Pickle 1998; Loiter and Norberg-Bohm 1999; Midttun and Kamfjord 1999; Norberg-Bohm 1999; Awerbush 2000; Jacobsson and Johnson 2000). The field grew livelier as actual policies multiplied and matured: it simultaneously informed and assessed policy evolutions and discussions.

The academic community got particularly involved in debates over the respective merits and flaws of price and quantity instruments. These began when the European Commission considered options for harmonisation in the late 1990s, but kept going in the following decades, especially since the Commission had not given up its ambition to institute a pan-European harmonised framework (e.g. Hvelplund 2001; Ménanteau et al. 2003; Lauber 2004; Jacobsson et al. 2009). Academic debates appear closely related to the status of renewable energy policy at the European and Member States level. In the case of the "FIT versus TGC" debate, while there were actual theoretical divergences, much of the disagreement resided on a "political basis" "around the role of government in supporting the RE sector" (Lipp 2007, p. 5422; interview, energy economist, 2013).

Theoretical and Practical Refinements in the Conception of Renewable Electricity Policy

The close relationship between research and policy, between theory and practice, shows in the evolution of debates around renewable energy support. These evolved from a relatively static approach to a more dynamic one: instead of identifying which types of instruments were the best (i.e. most effective, most efficient, least costly for society, and/or safest for investors) to develop RES-E capacity and production, the concern was the tuning of incentives to each stage of technology and market development. Such preoccupations are similar to those the EEG reform addressed. Indeed, German renewable energy policy influenced other European countries' policy choices. Having started to promote RES-E earlier than most, Germany was also first to face the consequences of its development; the issues Germany sought to address in 2000 thus gained importance as renewable energy progressed throughout the decade.

Besides the matter of evaluating external costs, three issues were central in discussions about RES-E policy in the 2000s: the need to take into account the diversity of renewable energy technologies, increased attention to cost histories and innovation trajectories, and the adjustment of instruments to these innovation trajectories.

Estimating External Costs

External costs were a key figure of the justification for renewable energy policy support in the framework of the liberalisation process. As we have shown in the previous chapters, the Commission justified policy intervention to support renewables as a way to correct some imperfections of the electricity market by approximating the internalisation of environmental and innovation externalities. In such an approach, the estimation of these external costs was crucial. Thus, since 1991, the Commission had supported the Extern-E project, a European Research Network working on the development of a common multidisciplinary methodology for estimating the external costs of energy production. The network conducted case studies throughout Europe and presented conclusions, some of which were published by the Commission's Directorate General for Research (DG Research) in 2003 (European Commission 2003). It suggested numbers for some external costs at both the EU and the national level. As stated in the DG Research's presentation, this estimation was proposed as an input for adjusting certain prices (such as electricity prices) or for informing decision-making processes (e.g. to set a carbon tax, a carbon price or RES-E support) so that these could reflect external costs without distorting market competition. Invoking external costs was thus associated with two potential virtues: one was to genuinely inform calculations; the other was to render RES-E support compatible with EU competition policy. As the DG Research recalls it:

> The recent Community guidelines on state aid for environmental protection explicitly foresee that EU member states may grant operating aid, calculated on the basis of the external costs avoided, to new plants producing renewable energy. (European Commission 2003, p. 5)

However, much of the academic literature published in the 2000s seemed to be rather focused on ways to adjust RES-E policy to the diversity of renewable energy technologies and to take innovation dynamics into account.

Taking the Diversity of Renewable Energy Technologies into Account
Support for RES-E mainly started with wind power, but by the mid-2000s renewable energies encompassed a much wider set of resources and technologies with very different cost profiles and so-called maturity.[2] Increasing attention was paid to the variety of RES-E generation technologies and to their specificities. As support was also intended to help bring renewable energy technologies out of the niches, views shifted from a one-size-fits-all policy towards instruments tailored to specific

technologies (Midttun and Koefoed 2003; Haas et al. 2004; Sandén and Azar 2005). For instance, Haas et al. stressed that:

> A mix of policy instruments needs to be *tailored to the particular renewable energy sources and the specific national situation* to promote the evolution of the renewable energy sources from niche to mass-markets. This policy mix needs to evolve with the technology. (Haas et al. 2004, p. 838, emphasis added)

With its technology-specific tariffs, the EEG was the first example of this approach, which required expertise on the technological and economic characteristics of various RES-E generation technologies.

Following Cost and Innovation Trajectories

As a corollary of this technology-specific approach, attention was increasingly paid to cost histories and innovation trajectories. Many articles addressed the cost reduction, innovation and learning dynamics of various renewable energy technologies and discussed the relevance of learning-curve and experience-curve approaches and of market support as opposed to R&D funding (e.g. Nemet 2006; Papineau 2006; Jamasb 2007; Shum and Watanab 2008; Foxon and Pearson 2008; Schilling and Esmundo 2009).

The study of innovation trajectories and learning curves stems from a specific literature in innovation economics that emerged in the mid-twentieth century (Wright 1936; Hirsch 1952; Arrow 1962; Alchian 1963 [as cited in Papineau 2006]). The learning-curve and experience-curve concepts originated there. In their most common form, these concepts "define the cost or price of a technology as a power function of learning source in cumulative form such as installed capacity, output, or labour" (Jamasb 2007, p. 54). In short, "they provide a simple quantitative relationship between cost and the cumulative production of a technology" (Papineau 2006, p. 1). This relationship is measured by a "learning factor" which corresponds to the cost reduction for a given cumulative output.

These concepts were originally devised in the context of manufacturing and mature industries (Jamasb 2007) and spread by the Boston Consulting Group in the 1960s as a tool for advice on competitive strategy (Papineau 2006, p. 2). Papineau explains that the learning-curve concept attracted newfound interest in the context of the development of climate change policies, but that its focus shifted from production planning and strategic management to "endogenous technical change and the use of reliable estimates of technological learning rates as inputs of energy forecasting models" (Papineau 2006, p. 2, citing McDonald and Schrattenholzen 2001). The idea that policy-driven market deployment will accelerate innovation and trigger the cost reductions needed to make RES-E "a normal industry without special treatment" (Lauber 2004, p. 1413) draws from this tradition.

Considerations of the learning curves and respective levels of "maturity" of renewable energy technologies gradually merged with discussions on instruments designs and assessments. One of the main criteria used to compare categories of instruments was their ability to speed up technological progress at minimal social cost.

Tailoring Support Schemes to Diverse Renewable Energy Sources and Technologies

As a consequence of technological progress being considered in a more dynamic way, the ability of renewable energy policy to catalyse innovation in its successive phases became an important aspect of instrument design. Instruments on both sides of traditional dichotomies such as technology-push versus demand-pull or quantity versus price were now considered not so much as competing but as complementary devices suited to different technologies at different stages in their evolution towards full market competitiveness. Increasingly, the subtlety of instrument design lay in the ability to implement the right instrument for the right technology at the right moment. FITs were usually considered best as the initial trigger, while quantity-based instruments were seen as best suited for technologies closer to maturity.

The literature discussed the compatibility of diverse support instruments to stages of technological development and encouraged a dynamic approach to policy mixes in which enforced instruments would evolve to accompany technological progress (e.g. Ménanteau et al. 2003; Midttun and Gautesen 2007; Finon 2008). A similar concern was expressed in the European Strategic Energy Technology Plan (Commission of European Communities 2007a) that the Commission launched in 2007:

> The essence of the European Strategic Energy Technology Plan (SET Plan) will be to match the most appropriate set of policy instruments to the needs of different technologies at different stages of the development and deployment cycle. The SET Plan must therefore embrace all aspects of technological

innovation, as well as the policy framework required to encourage business and the financial community to deliver and support the efficient and low-carbon technologies that will shape our common future. (Commission of the European Communities 2006, p. 8)

This concern for the fit between support instruments and technology "maturity" translated into attempts to make instruments themselves better able to adjust to the market and technology evolutions they were meant to accompany. This resulted in sophistication and refinement. For instance, technology bands were included into TGCs in order to make them able to differentiate between technologies and to avoid lock-in into the most competitive ones. FITs were supplemented with decrease mechanisms meant to adjust incentives to cost reductions. The aim was to enable instruments to take into account an increasing amount of information, so that they could follow technology and market developments as closely as possible.

Documenting the Experiment: Surveys, Assessment and Formalisation

The development of expertise and research on renewable energy policy was partly driven by the application of the 2001 Directive. In particular, the commitment by the Commission to monitor and assess implementation in Member States fed the constitution of a collection of evaluations of RES-E progress in EU Member States and of assessments of the increasing variety of support instruments and policies. The trend was reinforced after 2008 and the adoption of binding targets for Member States. As the IPCC Special Report on Renewable Energy Sources notes, "much of the literature describing and comparing these instruments, including their costs, is European and grey, stimulated largely by the need of EU countries to fulfil their Renewable Energy Directive requirements by 2020" (Mitchell et al. 2011, p. 45).

As planned by the Directive, the European Commission reports regularly on renewable energy progress within the EU (Commission of the European Communities 2004, 2005, 2008). To do so, besides Member States' national reports, the Commission used data, model outputs and assessments from expert reports that it commissioned and funded. Not only did these reports directly contribute to the build-up of knowledge and expertise on renewable energy support schemes, they were also partly carried out by academics. The projects assessing RES-E support have thus funded research in economics and public policy, and the evaluation of EU policy has fed academic literature on RES-E deployment.

In this light, the development of policy-supported RES-E markets in Europe in the 2000s could be likened to a process of live, scale-one experimentation: a variety of experiments were carried out and carefully documented and compared, innovations and hypotheses were tested both *in vitro* (using economic models) and *in vivo* (Callon and Muniesa 2003; Muniesa and Callon 2007; Callon 2009), all with the purpose of improving the effectiveness and efficiency of incentive schemes and, in the long run, developing sustainable renewable energy markets that are fully integrated to the European internal electricity market. At least, this is the picture that the rhetoric of the European Commission seems to paint.

The Commission, while sticking to its ambition to achieve EU-wide harmonisation in the long-term, started to present the coexistence of a wide diversity of support schemes in EU Member States as an asset. Given the lack of previous experience in supporting RES-E and the "immaturity" of existing support schemes and policies, in the Commission's discourse, policy diversity provided an opportunity to gather an impressive amount of know-how in the field, and to work towards better solutions in the process. This is pretty much what the European Commission argued in the annex to its 2005 Communication on the support for electricity from renewable energy sources:

> While gaining significant experience in the EU with renewables support schemes, competing national schemes could be seen as healthy at least over a transitional period. Competition among schemes should lead to a greater variety of solutions and also to benefits: for example, a TGC system gains from the existence of a feed-in tariff scheme, as the costs of less efficient technologies fall due to the technological learning, which in turn leads to lower transfer costs for consumers. Systems are already leaving behind the 'great divide' between price- and quantity-based approaches. This might be the way forward, with specific instruments aimed at specific policy goals and the overall support framework intelligently linked to other electricity market regulation. (Commission of the European Communities 2005, p. 16)

Then, to what conclusions did this process of trial-and-error lead, and did it modify the way FITs were considered?

The Legitimation of Feed-in Tariffs

The implementation of RES-E support combined with the development of expertise on renewable energy policy contributed to the validation of FITs at the EU level. By the mid-2000s, FITs were increasingly presented as the best option for promoting RES-E promotion. Both in the literature and in European Commission documents, most evaluations concluded that FITs overall proved more effective and less costly than TGCs (Commission of the European Communities 2005 as cited in Lipp 2007; Jacobsson et al. 2009; Couture and Gagnon 2010). This somewhat contradicted expectations that price-based mechanisms were more prone to capture and windfall profits, but FITs appeared more successful in directing investment towards RES-E generation and in driving rapid increases in RES-E installed capacity. The EEG, in particular, had impressive results on the deployment of renewable energy, especially wind and solar power (Jacobsson and Lauber 2006). Countries with quota-based systems, such as the UK or Sweden, did not display such straightforward results, in particular where less mature RES-E technologies were concerned (Jacobsson et al. 2009) and when considering overall costs. Lipp (2007 p. 5492) thus noted that

> Several recent policy assessments show, however, that the UK [Renewable energy obligation] as well as other European quotas systems produce renewable electricity at a higher cost than the FIT.

Couture and Gagnon (2010, p. 955) were even more straightforward:

> [FITs] have consistently delivered more renewable energy supply more effectively, and at lower cost, than alternative policy mechanisms.

In the following section, we review the main arguments that were put forward in favour of FITs. We suggest that they reveal a shift in the frame of reference for assessing RES-E policy from a conception of market support in terms of competition to one in terms of entrepreneurship and investment.

Feed-in Tariffs and Investment

The introduction of RES-E generation targets (albeit indicative) combined with an increased attention on learning processes in the literature on renewable energy made the increase in installed capacity a critical factor from both a static (achieving current objectives) and dynamic (setting RES-E on the path to competitiveness) perspectives. Increase in installed capacity is the result of the deployment of renewable energy technologies, which itself results from investments in renewable energy projects. There is a short distance from there to a concern for investment conditions. At a micro-level, RES-E policy can be considered as what is required to make investments in RES-E project attractive: it enhances their potential profitability and reduces project risks.

The ability of RES-E support schemes to provide guarantees for investment was increasingly considered as a key condition for market uptake. It was put forward as the main explanation for the success of FITs. Indeed, when it comes to reducing investment risk, few mechanisms fare better than FITs, which guarantee long-term visibility over future cash flows. This was reinforced by the fact that FIT levels were more and more often calculated with reference to the cost of RES-E generation, so in a fashion that directly connected them to an investor's perspective (Jacob 2012). As Couture and Gagnon explain:

> ... by basing the payment levels on the costs required to develop renewable energy projects, and guaranteeing the payment levels for the lifetime of the technology, FITs can significantly reduce the risk of investing in renewable energy technologies and thus create the conditions to rapid market growth (Lipp 2007; International Energy Agency 2008). This structure provides a high degree of security over future cash flows and enables investors to be remunerated according to the actual costs of renewable energy project development. This security is particularly valuable for financing capital-intensive projects with high upfront costs and a high ratio of fixed to variable costs. (Couture and Gagnon 2010, p. 956)

Quantity-based systems could not provide such certainty over return on investments, because they provided no real visibility on RES-E prices. As a result, their good functioning usually turned out to require payment of risk premiums (i.e. higher remuneration for the same level of production is usually required). This would explain why TGCs and bidding systems turned out to be more costly than FIT schemes on the whole.

> In theory, this difference should not exist as bidding prices that are set at the same level as feed-in tariffs should logically give rise to comparable capacities being installed. The discrepancy can be explained by the higher certainty of

current feed-in tariff schemes and the strong incentive effect of guaranteed prices. (Sims et al. 2007, as cited in Mitchell et al. 2011)

Similarly, Finon mentions the "stable signal quality that is specific to feed-in tariffs and that is favourable to investments in capitalistic equipment, as opposed to other schemes such as tradable certificate obligations, which do not offer reliable revenue" (Finon 2008, p. 15, authors' translation). The European Commission rallied to this view, noting in its 2011 review of European and national financing of renewable energy that

> reviewing the relationship between project risk and instrument choice, the empirical evidence suggests that the more reliable revenue stream provided by feed-in tariffs is generally more effective in driving renewable energy growth, particularly for a broad range of technologies. Quota obligation and tradable green certificates often suffer from revenue volatility and require payment of a risk premium, which appear to make them both less effective and efficient. (European Commission 2011, p. 6)

Feed-in Tariffs and Diversity

A second argument in favour of FITs had to do with the nature of RES-E generation projects themselves. FITs were shown to enable the emergence of a very diverse range of renewable energy technologies and projects, in large part because of their reassuring character for all kinds of investors. FITs do not directly expose RES-E projects to competition. Whereas TGCs support renewable energy in a generic manner[3] and imply that all kinds of projects are exposed to the same market conditions in order to direct subsidies towards the least costly solutions, FITs can be tailored to take into account the specificities of various technologies and investment models. The support they provide is not limited to the most mature technologies and the areas with the highest renewable energy potentials. As the Special Report on Renewable Energy (SRREN) noted, they "have encouraged both technological and geographical diversity, and have been found to be more suitable for promoting projects of varying size" (Mitchell et al. 2011, p. 55).

This can be considered a flaw in a pure "cost-efficiency" perspective, but advocates of FITs have often stressed their flexibility and their ability to generate diversity as key advantages (Hvelplund 2001; Lipp 2007; Couture and Gagnon 2010). On top of making it easier to promote renewable energy technologies in their diversity, FITs indeed allow non-traditional actors to enter the market: they enable "a greater number of investors to participate, including homeowners, landowners, farmers, municipalities, and small business owners, while helping to stimulate rapid renewable energy deployment in a wide variety of different technological classes" (Couture and Gagnon 2010, p. 955). Several analysts thus argued that FITs were suited to the characteristics of RES-E, because they were effective in "favouring early and rapid growth" of technologies still relatively far from market competitiveness (Lauber 2004, pp. 1411–1412) and with high upfront costs, while permitting their development at diverse scales and by various actors (Hvelplund 2001; Couture and Gagnon 2010).

On the contrary, TGCs have been shown to favour already powerful incumbents, which ironically would make them prone to generate excess profits and to facilitate regulatory capture by vested interests (Jacobsson et al. 2009). In a harsh critique of TGCs, Jacobsson et al. explain the persistence of the European Commission in pushing the "pan-European TGC dream" by a "neo-liberal ideology" and a coalition of vested interests, thus overturning the arguments that FITs are less trustworthy because they are dependent on politics. They point to the influence of neoclassical economics in the worldview of different General Directorates of the Commission and in the Commission's "preference for textbook analyses which are far away from the real world of complexity and uncertainty", as well as the "symbiotic relationship to the conventional power sector" maintained by regulators at the national and Commission levels. They also argue that TGCs allowed the Commission and, above all, power producers, to assert their power and control over the energy sector. First, TGCs made it easier to frame RES-E support as "falling within the competition and internal market remit", granting the Commission more power over energy issues by incorporating them in the Single Market agenda. Second, as incumbents benefit more easily from TGC systems and can capture excess profits because they are less exposed to risks, such systems "provide them with the market control and the political power to deploy renewables at their chosen pace" (Jacobsson et al. 2009, p. 2146). In this perspective, quota systems are no longer the market-based mechanisms eschewing the arbitrariness of politics that the Commission's narrative pictured them to be. On the contrary, they are presented as stemming from the politics of dominant players seeking to maintain their grip on the energy sector, whereas FITs are described as levelling the market, thereby welcoming new entrants and encouraging investments and entrepreneurship.

The European Commission Endorses Feed-in Tariffs

By the late 2000s, evidence suggested that FITs were the best instruments for the promotion of renewable energy. The German EEG was widely hailed as a success and held as an example (Interview, energy economist 2013; Mitchell et al. 2011, p. 52). The academic literature leaned towards similar conclusions, with articles stressing that "evidence suggests, albeit tentatively, that feed-in tariffs (FITs) are more effective than alternative support schemes in promoting renewable energy technologies" (Lesser and Su 2008, abstract), or that "recent experience from around the world suggests that feed-in tariffs (FITs) are the most effective policy to encourage the rapid and sustained deployment of renewable energy" (Couture and Gagnon 2010).

The European Commission eventually rallied around this position. Between 2001 and 2008, on top of monitoring the implementation of the 2001 Directive in Member States (Commission of the European Communities 2004, 2005, 2008), the Commission was working on the preparation of a second energy package that would seek, among other objectives, to further increase the use of renewable energy sources (Commission of the European Communities 2006, 2007a, b, c). In an evaluation of the support of RES-E that accompanied the proposal for a directive on the promotion of the use of energy from renewable energy sources, it admitted that:

[This report] finds that, as in 2005, well-adapted feed in tariffs regimes are generally the most efficient and effective support schemes for promoting renewable energy. (Commission of the European Communities 2008, p. 3)

While this conclusion was certainly informed by observations of the effects and costs of renewable energy policy in Member States, the Commission's argument stressed the incorporation of "market signals" in FIT schemes and their resulting better compatibility with internal market rules:

Several Member States have reformed their support schemes to differentiate between technologies to encourage technological diversity. Although the basic nature of the existing support schemes in place varies between Member States as does the level of support to different technologies, *there are clear signs that a degree of convergence of important properties of the policy measures is emerging.* Support schemes have also been *reformed to introduce market signals* through the incorporation of market prices using premiums rather than feed-in tariffs, thus *improving the compatibility of the support with internal market rules* and adjustments of tariffs to reflect decreasing production costs. This results in both an *improvement of the existing measures and a gradual increase in the effectiveness and efficiency of support to promote renewable electricity.* (Commission of the European Communities 2008, p. 14, emphasis added)

It also noted that the sophistication of RES-E support schemes had blurred the distinction between price and quantity instruments:

As a result of incorporating elements of different schemes, *the clear distinctions between the different support schemes are fading and their known problems are diminishing*: technology-specific obligations or green certificates can ensure that such regimes no longer develop only the current cheapest technology; greater use of feed in premiums can ensure that producers have stronger incentives to minimise costs. (Commission of the European Communities 2008, p. 14, emphasis added)[4]

This change in position might also relate to a shift in policy priorities at the EU level. Even though the Commission kept pushing for harmonisation in the preparation of the second energy package, climate change and security of supply gained importance over the completion of an internal electricity market that "proved rather intractable" (Lauber and Schenner 2011, p. 523). In addition, in 2007, the Commission once again had to let go of its pan-European TGC proposal in the face of opposition from the Council, especially from Spain and Germany, for the same motives as in 2001 (Laubert and Schenner 2011; Solorio and Bocquillon 2017). Reflecting this shift, the second package was in fact an "Energy-Climate Package".

Three documents in the "Energy Package" proposed by the Commission on January 10, 2007 were relevant to RES-E support:

- a communication entitled "An Energy Policy for Europe", which outlined the ambition to define a framework for a high energy efficiency and low carbon emission economy and insisted on the importance of defining a long-term vision for energy technologies (Commission of the European Communities 2007c);
- the launch of the "Strategic Energy Technologies" (SET) Plan, which focused on innovation in energy technologies, particularly

renewable energy technologies, and justified RES-E support instruments as a way to drive innovation (Commission of the European Communities 2007a);

- a "Roadmap for renewable energy sources" stressing the risk not to meet the 1997 renewable energy development objectives and proposing a target of 20% RES in EU energy consumption by 2020 (Commission of the European Communities 2007b).

The Commission simultaneously published a communication on "Limiting global climate change to 2 degrees Celsius: the way ahead for 2020 and beyond" (Commission of the European Communities 2007d). The Energy-Climate package that was eventually adopted in late 2008 confirmed the coupling of energy policy and climate policy that the simultaneous release of these documents had suggested. It included a Directive on the promotion of the use of energy from renewable sources setting binding national targets for energy produced from RES on the basis of the overall EU target (20% energy from RES in total energy consumption by 2020) (European Parliament and Council 2009).

CONCLUSION

Between 2001 and 2008, with a medium-term EU-wide target for the share of RES-E in electricity consumption but no harmonised policy approach, various RES-E policy schemes co-existed throughout the EU. The European Commission still hoped that its proposal for a pan-European TGC scheme would be adopted later. In the meantime, it attempted to coordinate EU renewable energy policy through regular reviews and assessments of the progression of RES-E in Member States. The Commission's commitment to deliver progress reports fed demand for expertise and research on the topic. From the Commission's perspective, this phase could appear as a moment of scale-one experimentation from which to draw lessons on RES-E policy design, though in practice Member States very much called the tune.

A majority of Member States opted for FITs as the centrepiece of their RES-E policies. However, compared to the situation in the 1990s, the agencement of FITs evolved. The way in which the design of FIT schemes took into account and articulated politics, market information, and technology characteristics changed. A first major change came from the European Court of Justice, which stated in 2001 that FITs did not constitute State aid and were not incompatible with the rules of the Common Market. The German EEG initiated a second evolution by introducing technology-specific FITs whose levels would decrease annually, to take into account decreases in technology costs. This effectively made FIT levels less dependent on the discretion of politics, by tuning them to information on technology and project costs: the calibration of FITs in the EEG was connected to economic dynamics at least as much as it was to political dynamics. The chapter showed that similar conceptions could be traced in the literature, which increasingly investigated the tailoring of support to the dynamics of technologies at different stage of deployment.

By coupling the calibration of FITs to economic parameters, investment or project costs became an important reference for their design. While the reference to "external costs" remained a crucial part of the political and legal justification for RES-E policy, the actual design of FITs more often than not relied on a characterisation of renewable energy technologies, on evaluations of their costs and on theorisations of technological innovation dynamics. To an extent, notwithstanding the Commission's continuous support to the actual assessment of external costs of energy, one could argue that this phase of experimentation with, and concretisation of, RES-E support schemes entailed a shift from a discussion of RES-E policy in terms of principles and in relations to abstract theories of markets to a consideration of the more mundane matters of actual market design.

These two evolutions combined with a shift in European policy orientations. Environmental and climate issues were gaining salience, and European energy policy was increasingly tied to European climate policy. FITs also appeared to work quite well where they had been adopted. The period between 2001 and 2008 thus led to a legitimation of FITs at the European level, with the Commission changing position to consider them as "market-based" instruments compatible with its wider objectives. This shift might however not amount to a full approval, as the next chapter will explore.

Notes

1. In a similar view, Owen (2006) distinguished three perspectives in research on the obstacles to the entry of renewable energy into the mainstream of the power sector: a research, innovation and deployment one (focused on the nature of innovation, on industry strategies and on learning processes),

a market barriers approach (a view of the adoption of new technologies as a market process, focused on the framework within which investors and consumers make decision) and a market transformation perspective (interested in the practical dimensions of market building and influencing actors' attitude and decisions).

2. The definition of "maturity" cannot be taken for granted; it is in fact quite complex, since it can incorporate technological, industrial, economic or political dimensions—most of which are not stabilised as such when considering emerging technologies, hence the inverted commas. For the purpose of this paragraph, we take it to refer to a comparison of the dependence on support: wind power, for instance, has been around for longer than photovoltaic, and its costs have decreased enough to reach a level comparable with those of conventional electricity. Being closer to "competitiveness", it can be considered as needing less support than other, more expensive and less widespread technologies for which deeper cost reductions are still expected.

3. At least in their most basic forms. More sophisticated models with "technology bands" have been developed in order to avoid lock-in of the most mature technologies.

4. The conclusions presented in these two paragraphs are drawn from the OPTRES report, an expertise report commissioned by the European Commission, and whose executive summary expresses the same ideas using similar wording (Ragwitz et al. 2007, p. 2).

REFERENCES

Alchian, Armen. 1963. Reliability of progress curves in airframe production. *Econometrica* 3: 679–693.

Arrow, Kenneth. 1962. The economic implications of learning-by-doing. *Review of Economic Studies* 29: 155–173.

Awerbuch, Shimon. 2000. Investing in photovoltaic: Risk, accounting and the value of new technology. *Energy Policy* 28: 1023–1035. https://doi.org/10.1016/S0301-4215(00)00089-6.

Bergek, Anna, and Staffan Jacobsson. 2010. Are tradable green certificates a cost-efficient policy driving technical change or a rent-generating machine? Lessons from Sweden 2003–2008. *Energy Policy* 38: 1255–1271. https://doi.org/10.1016/j.enpol.2009.11.001.

Callon, Michel. 2009. Civilizing markets: Carbon trading between *in vitro* and *in vivo* experiments. *Accounting, Organizations and Society* 34: 535–548. https://doi.org/10.1016/j.aos.2008.04.003.

Callon, Michel, and Fabian Muniesa. 2003. Les marchés économiques comme dispositifs collectifs de calcul. *Réseaux* 122: 189–233. https://doi.org/10.3917/res.122.0189.

Commission of the European Communities. 2004. *The share of renewable energy in the EU.* Commission report in accordance with Article 3 of Directive 2001/77/EC, evaluation of the effect of legislative instruments and other Community policies on the development of the contribution of renewable energy sources in the EU and proposals for concrete action. COM(2004) 366 final. Brussels, 26 May 2004.

———. 2005. *The support of electricity from renewable sources.* Communication from the European Commission. COM(2005) 627 final. Brussels, 07 December 2005.

———. 2006. *A European strategy for sustainable, competitive and secure energy.* Green Paper. COM(2006) 105 final. Brussels, 08 March 2006.

———. 2007a. *Towards a European strategic energy technologies plan.* Communication from the Commission to the Council, the European Parliament, The European Economic and Social Committee and the Committee of the Regions. COM(2006) 847 final. Brussels, 10 January 2007.

———. 2007b. *Renewable energy roadmap. Renewable energies in the 21st century: Building a more sustainable future.* Communication from the Commission to the Council and the European Parliament. COM(2006) 848 final. Brussels, 10 January 2007.

———. 2007c. *An energy policy for Europe.* Communication from the Commission to the European Council and the European Parliament. COM(2007) 1 final. Brussels, 10 January 2007.

———. 2007d. *Limiting global climate change to 2 degrees Celsius. The way ahead for 2020 and beyond.* Communication from the Commission to the Council, The European Parliament, The European Economic and Social Committee and the Committee of the Regions. COM(2007) 2 final. Brussels, 10 January 2007.

———. 2008. *The support of electricity from renewable energy sources.* Commission staff working document accompanying document to the Proposal for a directive of the European Parliament and of the Council on the promotion of the use of energy from renewable sources [COM(2008) 19 final], SEC(2008) 57, Brussels, 23 January 2008.

Couture, Toby, and Yves Gagnon. 2010. An analysis of feed-in tariffs remuneration models: Implications for renewable energy investment. *Energy Policy* 38: 955–965. https://doi.org/10.1016/j.enpol.2009.10.047.

Dinica, Valentina. 2006. Support systems for the diffusion of renewable energy technologies—An investor perspective. *Energy Policy* 34: 461–480. https://doi.org/10.1016/j.enpol.2004.06.014a.

European Commission. 2003. *External costs: Research results on socio-environmental damages due to electricity and transport.* DG Research Report: EUR 20198. Brussels.

———. 2011. *Review of European and national financing of renewable energy in accordance with Article 23(7) of Directive 2009/28/CE.* Commission staff working document. Accompanying document to the Communication from the Commission to the European Parliament and Council [COM(2011) 31 final]. SEC(2011) 131 final. Brussels, 31 January 2011. European Court of Justice, 2000

European Court of Justice. 2001. *PreussenElektra v. Schleswag*. Judgement of the Court, 13 March 2001, Case C-379/98.

———. 2000. Opinion of advocate general Jacobs delivered on 26 October 2000, Case C-379/98.

European Parliament and Council. 2009. Directive 28/2009/EC of the European Parliament and of the Council of 23 April 2009 on the promotion of the use of energy from renewable sources and amending and subsequently repelling Directives 2001/77/EC and 2003/30/EC. *Official Journal of the European Union* L 140: 16–62.

Evrard, Aurélien. 2010. *L'intégration des énergies renouvelables aux politiques publiques en Europe*. Ph.D. thesis, Sciences Po Paris, Paris.

Federal Ministry for the Environment, Nature Conservation and Nuclear Safety. 2000. Act on granting priority to renewable energy sources, Berlin. Also published 2001 in *Solar Energy Policy* 70 (6): 489–504.

Finon, Dominique. 2008. L'inadéquation du mode de subvention du photovoltaïque à sa maturité technologique. *Working Paper 2008–09*, Paris: CIRED.

Foxon, Tim, and Peter Pearson. 2008. Overcoming barriers to innovation and diffusion of cleaner technologies: Some features of a sustainable innovation policy regime. *Journal of Cleaner Production* 16S1: S148–S161. https://doi.org/10.1016/j.jclepro.2007.10.011.

Freeman, Chris. 1996. The greening of technology and models of innovation. *Technological Forecasting and Social Change* 53: 27–39. https://doi.org/10.1016/0040-1625(96)00060-1.

Haas, Reinhaard, Wolfgang Eichhammer, Claus Huber, Ole Langniss, Arturo Lorenzoni, Reinhard Madlener, Philippe Ménanteau, Poul Erik Morthorst, Alvaro Martins, Anna Oniszk, et al. 2004. How to promote renewable energy systems successfully and effectively. *Energy Policy* 32: 833–839. https://doi.org/10.1016/S0301-4215(02)00337-3.

Haas, Reinhaard, Anne Held, Dominique Finon, Niels I. Meyer, Arturo Lorenzoni, Ryan Wiser, and Ken-Ichiro Nishio. 2008. Promoting electricity from renewable energy sources—Lessons learned from the EU, U.S. and Japan. In *Competitive electricity markets*, ed. Fereidoon P. Sioshansi, 91–133. London: Elsevier.

Haas, Reinhaard, Gustav Resch, Christian Panzer, Sebastian Busch, Mario Ragwitz, and Anne Held. 2011. Efficiency and effectiveness of promotion systems for electricity generation from renewable energy sources—Lessons from EU countries. *Energy* 36: 2186–2193. https://doi.org/10.1016/j.rser.2010.11.015.

Hirsch, Werner Z. 1952. Manufacturing progress functions. *Review of Economics and Statistics* 34: 143–155.

Hvelplund, Frede. 2001. Political prices or political quantities? A comparison of renewable energy support systems. *New Energy* 5: 18–23.

International Energy Agency. 2008. *Deploying renewables: Principles for effective policies.* Paris: OECD/IEA.

Jacobs, David. 2012. *Renewable energy policy convergence in the EU. The evolution of feed-in tariffs in Germany, Spain and France.* Farnham: Ashgate.

Jacobsson, Staffan, and Anna Johnson. 2000. The diffusion of renewable energy technology: An analytical framework and key issues for research. *Energy Policy* 28 (9): 625–640.

Jacobsson, Staffan, and Volkmar Lauber. 2006. The politics and policy of energy system transformation—Explaining the German diffusion of renewable energy technology. *Energy Policy* 34: 256–276. https://doi.org/10.1016/j.enpol.2004.08.029.

Jacobsson, Staffan, Anne Bergek, Dominique Finon, Volkmar Lauber, Catherine Mitchell, David Toke, and Ariel Verbruggen. 2009. EU renewable energy support: Faith or facts? *Energy Policy* 37: 2143–2146. https://doi.org/10.1016/j.enpol.2009.02.043.

Jaffe, Adam B., and Robert N. Stavins. 1995. Dynamic incentives of environmental regulations: The effects of alternative policy instruments on technology diffusion. *Journal of Environmental Economics and Management* 29: S43–S63. https://doi.org/10.1006/jeem.1995.1060.

Jamasb, Tooraj. 2007. Technical change theory and learning curves: Patterns of progress in electricity generation technologies. *The Energy Journal* 28 (1): 51–71. https://doi.org/10.5547/ISSN0195-6574-EJ-Vol28-No3-4.

Lauber, Volkmar. 2004. REFIT and RPS: Options for a harmonised community framework. *Energy Policy* 32: 1405–1414. https://doi.org/10.1016/S0301-4215(03)00108-3.

Lauber, Volkmar, and Elisa Schenner. 2011. The struggle over support schemes for renewable electricity in the European Union: A discursive-institutionalist analysis. *Environmental Politics* 19: 127–141. https://doi.org/10.1080/09644016.2011.589578.

Lecuyer, Oskar, and Philippe Quirion. 2013. Can uncertainty justify overlapping policy instruments to mitigate emissions? *Ecological Economics* 93: 177–191. https://doi.org/10.2139/ssrn.2192517.

Lesser, Jonathan A., and Su Xuejuan. 2008. Design of an economically efficient feed-in structure for renewable energy development. *Energy Policy* 36: 981–990. https://doi.org/10.1016/j.enpol.2007.11.007.

Lipp, Judith. 2007. Lessons for effective renewable energy policies from Denmark, Germany and the United Kingdom. *Energy Policy* 35: 5481–5495. https://doi.org/10.1016/j.enpol.2007.05.015.

Loiter, Jeffrey M., and Vicki Norberg-Bohm. 1999. Technology policy and renewable energy: Public roles in the development of new energy technologies. *Energy Policy* 27: 85–97. https://doi.org/10.1016/S0301-4215(99)00013-0.

Lüthi, Sonja, and Rolf Wüstenhagen. 2012. The price of policy risk—Empirical insights from choice experiments with European photovoltaic project developers. *Energy Economics* 34: 1001–1011. https://doi.org/10.1016/j.eneco.2011.08.007.

McDonald, Alan, and Lea Schrattenholzer. 2001. Learning rates for energy technologies. *Energy Policy* 29: 255–261. https://doi.org/10.1016/S0301-4215(00)00122-1.

Ménanteau, Philippe, Dominique Finon, and Marie-Laure Lamy. 2003. Prices versus quantities: Choosing policies for promoting the development of renewable energy. *Energy Policy* 31: 799–812. https://doi.org/10.1016/S0301-4215(02)00133-7.

Midttun, Atle, and Kristian Gautesen. 2007. Feed in or certificate, competition or complementarity? Combining a static efficiency and a dynamic innovation perspective on the greening of the energy industry. *Energy Policy* 35: 1419–1422. https://doi.org/10.1016/j.enpol.2006.04.008.

Midttun, Atle, and Svein Kamfjord. 1999. Energy and environmental governance under ecological modernization: A comparative analysis of Nordic countries. *Public Administration* 77: 873–895. https://doi.org/10.1111/1467-9299.00184.

Midttun, Atle, and Anne Louise Koefoed. 2003. Greening of electricity in Europe: Challenges and developments. *Energy Policy* 31: 677–687. https://doi.org/10.1016/S0301-4215(02)00152-0.

Mitchell, Catherine, Janet Sawin, Govind R. Pokharel, Daniel Kammen, Zhongying Wang, Solomne Fifita, Mark Jaccard, Ole Langniss, Hugo Lucas, Alain Nadai, et al. 2011. Policy, financing and implementation. In *IPCC special report on renewable energy sources and climate change mitigation*, ed. Ottmar Edenhofer, Ramon PichsMadruga, Youba Sokona, Kristin Seyboth, Patrick Matschoss, Susanne Kadner, Timm Zwickel, Patrick Eickemeier, Gerrit Hansen, Steffen Schlömer, and Christoph von Stechow. Cambridge and New York, NY: Cambridge University Press.

Muniesa, Fabian, and Michel Callon. 2007. Economic experiments and the construction of markets. In *Do economists make markets? On the performativity of economics*, ed. David MacKenzie, Fabian Muniesa, and Lucia Siu. Princeton, NJ: Princeton University Press.

Nemet, Gregory F. 2006. Beyond the learning curve: Factors influencing cost reductions in photovoltaics. *Energy Policy* 34: 3218–3232. https://doi.org/10.1016/j.enpol.2005.06.020.

Norberg-Bohm, Vicki. 1999. Stimulating green technological innovation: An analysis of alternative policy mechanisms. *Policy Sciences* 32: 13–38. https://doi.org/10.1023/A:1004384913598.

Owen, Anthony D. 2006. Renewable energy: Externality costs as market barriers. *Energy Policy* 34: 632–642.

Papineau, Maya. 2006. An economic perspective on learning curves and dynamic economies in renewable energy technologies. *Energy Policy* 34: 422–432. https://doi.org/10.1016/j.enpol.2004.06.008.

Ragwitz, Mario, Anne Held, Gustav Resch, Thomas Faber, Reinhard Haas, Claus Huber, Poul Erik Morthorst, Stie Grenaa Jensen, Rogier Coenraads, Monique Voogt, Gemma Reece, Inga Konstantinaviciute and Bernhard Heyder. 2007. *Assessment and optimisation of renewable energy support schemes in the European electricity market.* OPTRES final report. Karlsruhe, February 2007.

Ringel, Marc. 2006. Fostering the use of renewable energies in the European Union: The race between feed-in tariffs and green certificates. *Renewable Energy* 31: 1–17. https://doi.org/10.1016/j.renene.2005.03.015.

Sandén, Björn A., and Christian Azar. 2005. Near-term technology policies or long-term climate targets? Economy-wide versus technology-specific approaches. *Energy Policy* 33: 1557–1576. https://doi.org/10.1016/j.enpol.2004.01.012.

Schilling, Melissa A., and Melissa Esmundo. 2009. Technology S-curves in renewable energy alternatives: An analysis and implication for industry and government. *Energy Policy* 37: 1767–1789. https://doi.org/10.1016/j.enpol.2009.01.004.

Schmalensee, Richard. 2012. Evaluating policies to increase electricity generation from renewable energy. *Review of Environmental Economics and Policy* 6: 45–64. https://doi.org/10.1093/reep/rer020.

Shum, Kwok L., and Chihiro Watanabe. 2008. Towards a local learning (innovation) model of solar photovoltaic deployment. *Energy Policy* 26: 508–521. https://doi.org/10.1016/j.enpol.2007.09.015.

Sims, Ralph E.H., Robert N. Schock, Anthony Adegbululgbe, Jørgen Fenhann, Inga Konstantinaviciute, William Moomaw, Hassan B. Nimir, Bernhard Schlamadinger, Julio Torres-Martinez, Clive Turner, et al. 2007. Energy supply. In *Climate Change 2007: Mitigation of climate change. Contribution of Working Group III to the Fourth Assessment Report of the Intergovernmental Panel on Climate Change*, ed. Bert Metz, Ogunlade Davidson, Peter Bosch, Rutu Dave, and Leo Meyer, 251–322. Cambridge: Cambridge University Press.

Solorio, Israel, and Pierre Bocquillon. 2017. EU renewable energy policy: A brief overview of its history and evolution. In *A guide to EU renewable energy policy: Comparing Europeanization and domestic policy change in EU member states*, ed. Israel Solorio and Helge Jörgens. Cheltenham: Edward Elgar Publishing.

Wiser, Ryan H., and Steven J. Pickle. 1998. Financing investments in renewable energy: The impacts of policy design. *Renewable and Sustainable Energy Reviews* 2: 361–386. https://doi.org/10.1016/S1364-0321(98)00007-0.

Wright, T.P. 1936. Factors affecting the cost of airplanes. *Journal of Aeronautical Sciences* 3 (4): 122–128.

CHAPTER 5

Turbulence and Reforms in European Renewable Energy Policy After 2008

Abstract The chapter retraces the period from 2008 to 2015 as a phase of many reforms and reconsiderations in renewable electricity policies throughout the European Union. Focusing on the example of photovoltaics, Cointe and Nadaï show that feed-in tariffs were re-opened and re-politicised along new lines. The authors outline several challenges in the design of feed-in tariffs as they were addressed and discussed in the literature. They then discuss the reception of these evolutions by the European Commission. The Commission eventually shunned feed-in tariffs and encouraged their phase-out, based on a concern for investment security and for the operation of the single market. This suggests the internal electricity market is now conceived primarily as a tool for the optimal orientation of investment.

Keywords Feed-in tariffs • European Union • Photovoltaics • Investors' confidence

The Commission's reports on the progress of electricity from renewable energy sources (RES-E) in the European Union (EU) may give the impression of a relatively controlled experimentation and review process allowing for learning and gradual sophistication and improvement of RES-E support instruments. A closer look at the evolution of RES-E policy since the late 2000s suggests a more intricate and turbulent picture.

© The Author(s) 2018
B. Cointe, A. Nadaï, *Feed-in tariffs in the European Union*, Progressive Energy Policy, https://doi.org/10.1007/978-3-319-76321-7_5

Focusing on the period between 2008 and 2015, this chapter explores the unexpected consequences of the success of feed-in tariffs (FIT) schemes. Despite their success in driving installed RES-E generation capacity, or perhaps because of this success, FITs soon raised new issues, in particular related to distributional effects and investors' confidence. We focus on the case of photovoltaics (PV), in which FITs turned out particularly difficult to manage in many countries, to analyse how the very effects of FITs reopened debates around the agencement of RES-E policy along new lines. The political dimension of FIT design came back to the fore, but in different terms than before. The challenges that arose in reforms of FIT schemes provide an opportunity to delve deeper into the technicalities of FIT agencing, and into their implications in terms of ordering the relationships among technology development, politics, and market dynamics. Having looked into these, we then turn our focus to the European Commission's reaction to these turbulences in RES-E policies and analyse the position and strategy it laid out between 2012 and 2014. The Commission seized the opportunity provided by the difficulties in managing FIT schemes throughout Europe to prompt a reconsideration of FITs.

The Evolution of Renewable Electricity Policies Across Europe in the late 2000s

Healthy Convergence?...

By the late 2000s, RES-E policies throughout Europe had matured, having been in operation for several years. Taken together, Member States had experimented on a variety of instrument types and designs that the multiplication of assessments, reviews, and studies had mapped out. This enabled the exchange of practices across countries, at least to an extent, potentially leading to policy convergence (Jacobs 2012 retraced the history of FITs asking whether they constituted an example of policy convergence across the EU). To reconcile multiple objectives, policy-makers invented hybrid systems along the way. These attempts at combining the advantages of price- and quantity-based instruments blurred the line between two theoretically well-established categories.

In its 2008 evaluation of the support of electricity from renewable sources, the European Commission greeted these evolutions with unabashed optimism:

Thus in general, it is clear that Member States are aware of and learning from the failing of their own support schemes and the best practices in other Member States. (Commission of the European Communities 2008, p. 14, emphasis added)

The Commission reiterated this stance in its review of Member States renewable energy policies released in 2011:

The use of multiple instruments or the adaptation of instruments also reflects Member States' efforts to improve the efficacy of the instruments in a gradual manner without causing too much disruption to the market. Changes in recent years have seen a blurring of the traditional dichotomy of tradable certificate (setting quantity not price) and feed in tariffs (setting price not quantity). [...] In addition, Member States make smaller annual changes—to the quotas, to the tariff or premium rates, to the lifetime of the support, and to aspects of eligibility. All of these changes improve the efficiency of the instruments. But more needs to be done. (European Commission 2011b, p. 7, emphasis added)

Such discourses presented the evolution of renewable energy policy in Europe as a coordinated experimental process. They pictured RES-E support schemes as well-defined, comparable devices that could be extracted from the specificities of any national economic and political contexts to be transferred in another country. In such a view, lessons learned from one country could be easily used in another, and so could improvements to policy instruments. Diversity across Member States was a good thing as long as it allowed for an exchange of "best practices".

A closer look at the evolution of renewable energy policy and at the literature documenting it suggests that the process did not go as smoothly as the Commission reviews state. In fact, the gradual development of RES-E generated new issues and overflows. Some were direct effects of flaws in the design of support schemes, and others resulted from difficulties to keep track of the dynamics of technological and economic change (see for instance Hoppmann et al. 2014 for an account of the evolution of German photovoltaic policy). The financial crisis and its budgetary consequences, combined with rapid decreases in the cost of RES technologies, also made the cost of RES-E support a more salient issue (Solorio and Bocquillon 2017). In most cases, the need to reform, adjust, and refine support schemes stemmed from a realisation that RES-E development was a more complicated problem than anticipated, or rather, that it brought new problems that required quick solutions. In short, as renewable energy policy matured, it turned from an *a priori* simple problem with a limited range of simple and elegant solutions to a messy process of bricolage and experimenting, as an official from a large French utility explained, referring in particular to the evolution of support for photovoltaics:

> One could say that, in fact, the variety of renewable energy support systems is limited, but experience has actually shown that it diversified all the same. [...]
> When the photovoltaic crisis began, we started trying to refine all the systems beyond this overly simple opposition. And that is where I think it was difficult, because we had to invent things that could make the incompatible compatible along the way. [...] So improvements were made on the tradable quota systems, one may say, but that at the same time they falsified it. On the tariff side, we said: 'we can see that it is drifting away', so experiments were tried such as the ones in Spain and Germany, either hybrid systems or systems that turned the logic upside-down. In the end, the difficulty was that we thought it was an extremely simple economic theory problem with an extremely limited number of solutions, and then we kept on refining it, trying to turn a few screws to get something that would combine too many objectives: we want to promote all technologies, we don't want tax to increase too much, we want a risk-structure that is acceptable for investors otherwise they will legitimately ask for better capital remuneration... There was this whole series of compromises, so we kept on patching innovations together. [...]
> To put it shortly, it took place in a great mess because we were in the thick of several problems: on the one hand, in some countries it was too expensive, and on the other hand, in some sectors we had trouble obtaining as smooth a development as many governments wanted. And in the end, we did tamper a lot with it... So people went on missions in neighbouring countries to see if Peter's solution could be transposed to Paul, if necessary with a few modifications. (Interview, senior official in electricity utility 2012)

... or Ad-Hoc Tinkering?

Rather than a sign of healthy convergence through experimentation and exchange of best practices, the gradual sophistication of RES-E policies appears as the result of a series of ad-hoc reforms made necessary by the need to adjust instruments to changing objectives and constraints. FIT schemes started to raise concerns around 2008, especially those supporting electricity generated from photovoltaics.

In 2008, Spain's photovoltaic support system collapsed. Following a blistering growth that had propelled Spain in the top-3 countries globally in terms of photovoltaic capacity but weighed heavily on the public budget, the 1758/2008 Royal Decree retroactively lowered tariffs and refined their design to curtail development. In Germany, where objectives were overshot on a regular basis, a revision of the *Erneuerbare-Energien Gesetz* (EEG) in 2009 introduced a system of dynamic FIT decrease based on the rate of development—the decrease rate had to be adjusted every year since then (Hoppmann et al. 2014). In the Czech Republic, a moratorium on FITs was decided in 2010, and a decrease in FIT rates came into effect the following year. In the UK, where FITs were introduced in 2010, those for photovoltaic power were cut and reviewed in late 2011. In France, the FIT system was reformed several times in 2010 with so little effect that a moratorium and thorough revision was eventually decided upon in December 2010 (Cointe 2017).

These reforms may have contributed to improvements in the design and efficiency of support schemes, as a report on renewable energy progress commissioned by the European Commission, suggests:

> Apart from those countries which have put their support for new installations on hold, most Member States are continuously refining and [sic] their support systems to improve their effectiveness and efficiency. (Harmelink et al. 2012, p. 102)

However, their relative brutality also generated unpredictability in a context in which reliability and investor confidence are crucial. The report stresses the joint influence of the financial crisis and the swift reduction of photovoltaic module costs in this matter, noting that

> Overall, we can observe that the recent economic crisis has affected the reliability of RES support in a number of member states. (Harmelink et al. 2012, p. 100)

and that

> A number of countries made abrupt changes to their RES-E support schemes in 2010 and 2011 to keep up with the rapid price development on the PV market (e.g. Spain, Czech Republic, the UK, Latvia, Portugal), but these changes undermined the confidence of the investors which is a serious threat to the success of RES policies in the future. (Harmelink et al. 2012, p. 12)

In fact, the effects of the financial crisis on renewable energy policy were twofold. First, renewable energy projects became particularly attractive investment options, because they were supported by mechanisms that were designed specifically to reduce investment risk—or, in the case of FITs, to effectively suppress risk. This was especially true for photovoltaic projects for two main reasons: they are relatively easy to carry out and, between 2008 and 2009, the cost of photovoltaic modules decreased dramatically (Bazilian et al. 2013), leading to discrepancies between project costs and FIT remuneration levels. This was a recipe for success and disaster at the same time: success was achieved in terms of installed capacity development, but its unexpected level translated into unexpected costs. A second consequence of the financial and subsequent economic crisis was the increased attention paid to the cost of renewable energy policy as well as to their economic, social, and industrial benefits (interview, senior official in electricity utility 2012). Renewable energy support was growing increasingly expensive while control of spending was becoming a major concern of European governments. The success of FITs in terms of increase in installed capacity triggered instability in support (Harmelink et al. 2012, p. 32).

From 2008 onwards, renewable energy policies or, at least, FIT schemes, went through turbulences and reforms. This fed a process of experimentation and bricolage to adjust support and take into account its economic and political costs. As a result, this re-opened the issue of instrument design in the light of new considerations. The problems of external costs and level competition in a liberalised internal electricity market seemed far away.

Critiques of Feed-in Tariffs

Critiques of renewable energy support started to appear in the literature published in the late 2000s. They highlighted some of the problems that had emerged from the implementation of renewable energy policy in several countries, often stressing that the development of RES-E could not be reduced to growth in installed capacity. This led to contestations of the adequacy of FITs in certain cases. The focus on instruments and on their effects in terms of installed capacity, as if RES could be considered only in terms of their tradable electricity output, may have drawn attention away from other crucial aspects; the literature was starting to look into these.

The Hegemony of Economic Instruments Reconsidered

The literature increasingly considered renewable energy policy beyond the sole assessment of specific instruments. Dinica, for instance, argued that the focus on one single type of instrument had led to overlooking the effects and importance of other policies that in some cases had been crucial. In a 2008 paper, she warned against a dominant "narrow conceptualization of policy referring mostly to direct instruments for economic feasibility" which "often led to unsatisfactory explanation of diffusion results" (Dinica 2008, abstract). The focus on a few instruments has obscured the complexity of renewable energy policy, she argued, since it often has led to the conclusion that "the described instruments led to the observed diffusion results, as if there was little to nothing in between" (Dinica 2008, p. 3563).

Taking the example of wind power development in Spain, Dinica showed that it owed much more to a political and institutional context that encouraged and facilitated public-private partnerships (PPPs) for wind power projects than to FITs themselves. Even though FITs were instituted in Spain in 1994, they provided a rather insecure framework for investments until 2004. They did not constitute much of an incentive for risk-adverse Spanish economic actors. PPPs, however, created the trust in wind power projects that was necessary to draw banks, developers, and industry in, and to generate a sustained growth in wind installations.

Dinica also regretted that the trope towards economic instruments that dominated in the academic and policy arena had led many to attribute the successful development of wind power in Spain to feed-in tariffs without thorough consideration of the evolution of policies and of diffusion patterns. The fact that feed-in tariffs were a successful trigger for renewable energy development in Denmark or Germany should not lead to the conclusion that they *always* are the main drivers for renewable energy development.

By shifting attention from debates on the design of a few economic instruments considered as the main elements of renewable energy policy, Dinica's paper brought the complexity of policy and market arrangements to the foreground. It emphasised that renewable energy policy could not be reduced to the details of instrument design, and that instruments were not immediately transferable from country to country without paying attention to specific political and economic contexts that largely shaped the way market players approached investment risks and innovation.

This view may not have become dominant, but it nonetheless altered the debate by drawing attention to the fact that there was more to renewable energy policy than instrument design. It was also a sign that the focus of researchers and policy-makers was shifting from an abstract, uniform conception of instruments to one in which renewable energy policy instruments had to be adjusted to a given market, for a given technology, in a given country, at a given time.

The Adequacy of Feed-in Tariffs for All Technologies Contested

Indeed, renewable energy support instruments were no longer considered in general terms: the issue was to determine their suitability for supporting the development of specific renewable energy technologies. The adequacy of FITs as the main support instrument was particularly debated in the case of photovoltaics. FITs had originally emerged to support wind power, and their suitability to other renewable energy sources with very different characteristics could not be taken for granted.

Some of the critiques of the application of FITs to photovoltaics stemmed from innovation studies. They questioned the relevance for photovoltaics of the "learning-by-doing" approach underlying RES-E policy. For instance, Finon (2008) argued that photovoltaic technologies were not yet mature enough to justify support through generation subsidies. Schilling and Esmundo (2009) took a similar stance: they chose to focus on "learning by searching" rather than "learning by doing" and showed that the amounts spent on photovoltaic support were out of proportion with their impact in terms of cost reduction and efficiency gain.

Many analysts have stressed that, even though the learning curve model seems to apply rather well to photovoltaic panels manufacturing[1] (van der Zwaan and Rabl 2004, p. 7), the relevance of learning-by-doing to evolving and emerging technologies such as renewable energy technologies remains uncertain (Jamasb 2007). Schmalensee (2012, p. 48) pointed out that learning "only provides an economic justification for subsidies if there are spill overs from one producer to others". In the case of photovoltaics, he wrote, such spill overs have not been demonstrated, and neither has it been shown that

> costs are more effectively reduced by subsidizing deployment of today's expensive technologies than by directly supporting research and development aimed at finding lower costs alternatives or offering prices tied to generation costs. (Schmalensee 2012, p. 48)

In fact, as van der Zwaan and Rabl (2004, p. 9) point out,

> [The learning curve] provides little to no explanatory value. This property of the learning curve methodology implies that it remains difficult to assess how precisely one needs to go about promoting PV or stimulating cost reduction.

Other critics dismissed FITs altogether as a means to support the deployment of photovoltaics, looking at their broader impacts on industries and employment. For instance, Frondel et al. (2008, 2010) questioned the apparent consensus over the "German renewable energy success". They argued that in spite of their effectiveness in promoting photovoltaic power, German feed-in tariffs had failed to bring about any of the benefits that could be expected from the deployment of photovoltaics. According to them, FITs had not proven to be able to help the emergence of domestic equipment industries and to create jobs. The costs of feed-in tariffs for PV-generated electricity, they write, were disproportionate, and photovoltaic power benefited from a preferential treatment that failed to contribute significantly to climate protection and/or job creation. This critique gained ground as the Chinese photovoltaic panels industry developed and flooded the European market, and as the financial crisis made the costs of renewable energy policy a critical matter.

CHALLENGES OF FIT DESIGN: THE EXAMPLE OF PHOTOVOLTAICS

FITs were criticised, but not for being insufficiently market-based, too dependent on arbitrary political decisions, or incompatible with competition law. New issues and expectations came into focus, such as industrial effects, employment, social cost, or the specificities of national political and economic contexts. This translated into the reforms of RES-E support schemes of the time, and created new challenges for FIT design.

Feed-in tariffs had to be adjusted to suit redefined priorities and adapted to a changing context. Debates about renewable energy policy no longer focused on the virtues of ideal types of instruments, especially since the increased hybridisation of price and quantity instruments made such discussion less relevant. Instead, increased attention was paid to details in their designs. As Lüthi and Wüstenhagen wrote,

one way of summarizing the debate is that for many renewable energy policies, the devil is actually in the details (Ringel 2006), and it is a fine-tuned
set of ingredients of a country policy mix rather than any archetype of a
'price-driven' or 'quantity-driven' policy instrument that results in efficient
and effective deployment of renewables (Dinica 2006; Ménanteau et al.
2003). (Lüthi and Wüstenhagen 2012, p. 1002)

As an example of this evolution, we focus on the design challenges that
the management of FITs for photovoltaics raised. Around 2010, the main
problem with FIT-supported photovoltaic markets was that they were
growing too fast. The success of feed-in tariffs in increasing installed
capacity led to overflows that turned out to be difficult to control. As a
result, the calibration of FITs became a crucial issue: the incentive provided by FITs needed to be strong enough to drive investments, but not
so strong as to over-stimulate the market.

There are two aspects to the calibration of feed-in tariffs. First, policy-
makers have to determine the initial feed-in rate; this first step is far from
straightforward, which was part of the justification for the European
Commission's preference for quantity-based instruments in the 1990s.
The evolutions of photovoltaic markets over the 2000s revealed an
additional challenge in calibrating feed-in tariffs: not only must prices be
set at the right level, they must also be kept at the right level. In other
words, mechanisms have to be devised to enable feed-in tariffs to adjust to
dynamic market conditions without damaging the reliability of support. In
terms of frames of references for price-setting, this is another evolution: in
discourses about RES-E policy, the reference to project costs that had
become dominant as a result of an increasingly investment-centred conception of FITs gave way to a conception of FIT levels as the result of
compromises established to manage multiple political and economic
uncertainties. However, here again, the matter was simultaneously
addressed in theory and in practice, and the theory fed on the practical
experiences of various countries.

What Is a 'Right' Tariff?

The first difficulty in FIT design is to set the price. From an economist's
perspective, FIT schemes require that policy-makers "substitute their
judgement for that of markets in the selection of long-term, technological 'winners and losers'" and "define administratively FITs attributes,

specifically payment amounts for individual technologies [...], payment structures (e.g. fixed or declining), and payment duration", all of which "can require significant 'guesswork' [...] as to future market conditions and rates of technological improvements" (Lesser and Su 2008, p. 982). Though these difficulties were identified and studied well before the late 2000s, determining the best-suited method and obtaining reliable information to decide upon the appropriate price level has remained a major challenge for policy-makers. Multiple references have been used throughout the years, from avoided costs to external costs to project development costs; each stemmed from different conceptions of RES-E policy and of its relations to the ordering of energy systems, markets, and politics.

Theoretical Approaches to Feed-in Tariff Calibration

Ménanteau et al. (2003 pp. 56–57) remark that, as "it is impossible to refer to an optimum level of renewable energy production", "one is forced to adopt a strict cost/efficiency approach in which the target is defined exogenously by political decision-makers on the basis of available scientific information, but without strict economic rationalisation". However, the "scientific information" that can help set tariff levels is notoriously uncertain.

The IPCC Special Report on Renewable Energy (SRREN) lists four main approaches to setting FIT levels: payments based on the levelled cost of renewable electricity generation (LCORE); payments based on the value of renewable electricity generation; payments determined through auction mechanisms; and "simple fixed-price incentives based on neither generation costs nor notion of value" (Mitchell et al. 2011, p. 50). In theory, the level value of renewable electricity generation could also be determined on the basis of its environmental benefits— this is even how it should be done according to French law (Loi no 2000-108, article 2). But this is hardly a practicable basis (Haas et al. 2008, p. 8; interview, French civil servant 1 2012). In practice, FIT levels are usually based on estimated production costs relative to electricity price (Haas et al. 2008, p. 8). Relying on "specific generation costs" and designing FITs "to make it possible for efficiently operated renewable energy installations to be cost-effectively installed" is often considered the most effective and practicable option (Couture and Gagnon 2010, p. 955).

Elusive Information on Project Costs

Reliable information on the current and projected cost of renewable energy technologies and projects is not simple to obtain—especially for photovoltaics, considering the dramatic evolutions in the prices of photovoltaic modules since 2008 (Bazilian et al. 2013).

The information necessary to determine FIT payment is not only difficult to track; policy-makers often depend on industrial actors and projects developers to obtain it. These stakeholders may have interest in high FIT payments, as the renewable energy policy and economics literature stressed. Finon (2008, pp. 15–16, authors' translation) outlines that "the influence of interest groups and industrial actors who want to develop in this technological domain is noticeable on all the characters of the mechanisms: level, duration, sequential evolution of price during the contract, price decrease from one year to the next". Lesser and Su (2008, p. 986) also stress the challenge "to efficiently elicit truthful information for this industry without undue administrative burden" in a situation where "the right information set is fundamental to the effectiveness of a FIT structure".

This dependence puts regulators in a difficult position, as they not only have to guess which FIT level is appropriate, but also to justify why they do not rely entirely on the data provided by stakeholders. A civil servant in the French administration explained:

> Basically, how do we calculate tariffs? We're a bit bereft, because we have data on the decrease of panel, well, of photovoltaic modules, photovoltaic cells, but we never have... no industrial or project developer is ever going to tell us: 'my project works at 1 € per Wp installed, or 2 € per Wp installed'. People will never tell us that, or if they do, that's not the numbers they have, that's other numbers. And we can estimate, but we're never precise at 20 or 30%. So at some point there's a phase where we suggest something, see the reactions, and then we adjust, there's a phase of discussions-negotiations. That's a shame, but that's the way it is in many sectors. People like to think that the administration is able to perform expert calculations and say: 'well, actually the tariff must be like that', that's... no. (Interview, French civil servant 2 2012)

Ultimately, the establishment of FIT levels depends on the outcome of political negotiations, in particular on the interests and relative influence of the stakeholders in renewable energy policy (Cointe 2017). This became increasingly apparent with the difficulties in adjusting, reforming, and maintaining FIT schemes.

Adjusting Feed-in Tariffs

Keeping Up with the Dynamics of Photovoltaic Technologies Markets

The difficulty in obtaining reliable information on costs was heightened by the quick evolution in the costs of renewable electricity technologies and production, which the level of support partly influenced. This implied that FITs had to be (almost constantly) adjusted so as to follow technological and economic evolutions. In the case of photovoltaics, these have been particularly quick and hard to predict, especially as the combination of high generation subsidies in Europe and manufacturing subsidies in China has triggered the expansion of the Chinese photovoltaic modules industry, resulting in drops in photovoltaic module prices (Jäger-Waldau 2013). The dynamism of the global photovoltaic industry virtually left no time for the "trial and error process" through which FIT levels would be adjusted in theory (Ringel 2006, p. 9).

In such a context, the stability of feed-in tariffs started to be viewed as "also an Achilles' heel: a fixed, long-term price—or a price series with a built-in technology adjustment factor—will almost certainly deviate from realized market prices by greater amounts over time, thus distorting wholesale and retail energy markets" (Lesser and Su 2008, p. 983).

However, adjusting FIT levels is a complicated matter, especially when it has to be done often. When decided by policy-makers to follow changing market trends, FIT adjustments indeed become hard to predict and can lead to "stop & go". This has been shown to generate risk and uncertainty and undermine investors' confidence, and so to be counterproductive in terms of market support (Mitchell et al. 2011). As the IPCC Special Report on Renewable Energy summarised:

> the higher the frequency of adjustments [...] and the higher the degression rate in case of overshoot, the greater the control of support but the lower the stability for investors. (Mitchell et al. 2011, p. 52)

Stepped FIT such as those introduced by the German EEG or such as the French self-adjusting FITs were devised precisely to meet this challenge, and are examples of "design features" that have been "inherently built into the schemes to ensure they are flexible enough to account for changes in the development of costs and technologies and so minimise the financial support granted" (European Commission 2013a, p. 3). However, they have not been sufficient to spare the need for reform: combining stability and flexibility remains a challenge with no pre-printed solution.

Avoiding Over-Stimulation

Mismatches between policy incentives and actual market conditions became a problem because they entailed under- or, more frequently, over-payment and over-stimulation of the market (Mitchell et al. 2011, p. 52). The latter has been the main problem with photovoltaic support in Europe since the late 2000s, and support scheme design came to focus on ways to enhance control of RES-E growth.

Excessive feed-in tariffs led to over-stimulation of the market because they provided over-payment to photovoltaic electricity producers; since the cost of FIT policies was borne collectively (usually by electricity users), the issue of controlling market growth emerged mainly as a distributional issue. This re-opened the political dimensions of renewable energy support to contestation and discussions, because it raised the question of how much it is fair to pay collectively to develop RES-E generation, and of how it should be balanced with other objectives. Feed-in tariffs have thus been criticised for weighing rather heavily and lastingly on electricity prices and for amounting to a form of tax:

> Concerns about distributional impacts of renewable energy policies on poorer consumers arise most frequently in countries where FITs have led to significant increase in RE capacity, particularly for relatively high cost technologies such as PV, because of resulting increase in total electricity costs. (Mitchell et al. 2011, p. 58)

> Even the highly successful German FIT—successful when measured in terms of renewable capacity developed—has been criticised for its adverse impact of electric rates and retail customers increasingly protest its implementation. (Lesser and Su 2008, p. 982)

The design of compensation mechanisms (in France and Germany, a levy on electricity consumption) became a crucial and highly sensitive issue.

In fact, the more secure the investment conditions a FIT scheme provided, the higher the collective risk of seeing the market deviate and lead to high costs. Schmalensee insisted on this characteristic of FIT systems, shedding new light on their supposedly risk-free nature:

> The clearest theoretical argument for FIT's superiority over RPS is that guaranteeing the price at which a renewable generator can sell removes electricity market risk from investors in renewable generation, so that more capital can

be raised per dollar of subsidy expense. But this bang for the buck measure neglects the impact on actors other than investors in renewables and those who pay subsidies. Measures that remove market risk from one set of players may simply shift it to others and thus not reduce the risk to society as a whole. (Schmalensee 2012, p. 50)

While he regretted that "overall social risk seems to have received little attention from analysts", Schmalensee noted that simple design features could reduce the riskiness of policy measures for society—for instance a cap on the maximum quantity of electricity that can benefit from the guaranteed price. Such measures, however, reduce investment security and visibility and can even be harmful to market development; so it is a difficult balance to strike. Renewable energy policy was then no longer conceived just as a liberalisation issue, nor as a mere investment issue, but also—and, sometimes mainly—as a collective problem.

Balancing Private and Collective Risks and Benefits
Since the 2008–2010 FIT crises, finding and maintaining a balance between investor security and social cost has become an increasingly important focus of photovoltaic support policy. To an extent, risk reduction for everyone (government, investors, electricity consumers, and taxpayers), or rather, the quest for as equitable an allocation of risks as possible, has become a determining criteria in designing and amending RES-E policies.

The swift increase in renewable electricity generation capacity in the mid to late 2000s reassured the EU about the capacity of Member States to meet their 2010 and 2020 targets (European Commission 2011a, b), but progress then slowed down. European countries were successful in achieving rapid growth in renewable electricity generation, but the very conditions of this success, that is incentives that shifted the burden of uncertainty from investors to governments and/or energy consumers, put their sustainability into question.

Such concerns translated in the design of FIT schemes. They led to attempts to control the quantity and quality of projects able to benefit from feed-in tariffs. In terms of quantity, stepped FITs, such as those used in Germany or in France after 2011, were meant to limit the costs of feed-in tariff schemes; in France, they explicitly included a cap on the annual volume of installation, according to which the evolution of FIT rates was calibrated. In terms of quality, these distributional issues led policy-makers to refine FIT schemes in order to target market development to sectors deemed more equitable, more promising for society as a whole, or less speculative. France, in particular, went into a great degree of sophistication in differentiating support according to the type and size of photovoltaic projects: for instance, smaller, building-integrated installations benefitted from FITs, whereas large ground-mounted projects became allocated via a tendering procedure.

Distributional issues were even heightened by the realisation that feed-in tariffs had not been so successful in triggering industry deployment and job development in Europe. Even in Germany, the industrial benefits of photovoltaic support have been contested (Frondel et al. 2008). Because of the expansion of the Chinese photovoltaic manufacturing industry, cheap imported panels flooded European markets, undermining the emerging industry. FIT systems were clearly more successful in supporting photovoltaic electricity producers and project developers than in developing photovoltaic components manufacture. In France, feed-in tariffs were blamed for subsidising the Chinese industry. To remediate this and turn FITs into mechanisms for supporting European industries, Italy and France temporarily introduced premiums for photovoltaic installations that included components produced in the EU, though these were eventually rejected by the WTO.

Several of these reforms of FIT schemes negatively affected investment. The challenge was not only to allocate the cost of the development of photovoltaics in a more acceptable way, but also to sustain growth while guaranteeing that it occurred in a politically reliable and economically viable manner. Since the turbulences and reforms that occurred around 2010 seriously damaged investors' confidence, they ensured that it remained a crucial issue: restoring a reasonable level of security for investors while avoiding overshoot became the crux of RES-E policy. This concern also translated into increased pressure from the Commission for Member States to reduce non-economic barriers as an alternative way to reduce risks:

> [...] administrative burdens and delays still cause problems and raise project risk for renewable energy projects; slow infrastructure development, delays in connection, and grid operation rules that disadvantage renewable energy producers all continue and all need to be addressed by Member States in the implementation of the renewable energy Directive. (European Commission 2013b, p. 32)

THE COMMISSION'S NEW APPROACH TO RENEWABLE ENERGY POLICY

When reviewing renewable energy progress in 2013, the European Commission stressed the issue of investor confidence. While it noted the "negative impact on investment in renewable energy" of the financial crisis and its influence on the financial risk rating of EU countries (European Commission 2013b, p. 6), the Commission laid a significant part of the blame on what it considered a lack of anticipation and a bad handling of reform by Member States—or, in its own terms, on "the increase in risk resulting from Member States changes to support scheme" (European Commission 2013b, p. 32). It stressed that "as investor and market confidence in the renewable energy sector depends heavily on the regulatory framework, the reform of support mechanisms must be managed carefully" (European Commission 2013b, p. 9). Support schemes, the Commission stated in another communication in 2013, "should not frustrate investor's legitimate expectations" (European Commission 2013d, p. 15).

In an earlier communication entitled "Renewable Energy: a major player in the European energy market", the Commission's criticism was even more explicit. It painted a very different picture from that of experimentation, learning, exchange of best practices, and healthy convergence that the Commission had used to characterise the earlier development of a diversity of approach to RES-E policies across Europe. Now, not only did the way Member States handled RES-E policy "undermine investor confidence", their diverging approaches to RES-E support risked "impairing the single market":

> Recent changes to support schemes have in some cases been triggered by unexpectedly high growth rapidly increasing expenditure on renewable energy which is not sustainable in the short term. In some Member States, changes to support schemes have lacked transparency, have been introduced suddenly and at times have even been imposed retroactively or have introduced moratoriums. For new technologies and investment still dependant (sic) on support, such practices undermine investor confidence in the sector. Moreover diverging national support schemes, based on differing incentives may create barriers to entry and prevent market operators from deploying cross-border business models, possibly hindering business development. Such a risk of impairing the single market must be avoided and more action is also needed to ensure consistency of approach across Member States, to remove distortions and develop renewable energy resources effectively. (European Commission 2012a, p. 5)

Despite a focus on investor's confidence—which, the Commission argued, would be best guaranteed by a well-functioning internal market—the Commission also recognised that a growing array of objectives had to be taken into account when calibrating RES-E support. Public intervention needed to be adjusted "in order to stimulate innovation, increasingly expose renewables to market prices, prevent overcompensation, diminish the costs of support and ultimately end support" (European Commission 2013d, p. 3, referring to European Commission 2012a). The pragmatics of RES-E policy, as viewed by the Commission, had evolved, but they remained subordinated to good integration of RES-E in the internal electricity market.

Coordinating Support Through Guidance and Guidelines

The Commission developed a new approach to reassert this subordination of RES-E policy to the framework of the internal electricity market. EU-wide harmonisation of renewable energy policy had not worked. It had even been firmly opposed by some Member States, chiefly Germany and Spain (Solorio and Bocquillon 2017). The 2030 climate and energy framework adopted in 2014 to an extent reduced the EU's influence in favour of a renationalisation of renewable energy policy, since it did not include binding national targets (Solorio and Bocquillon 2017). At the same time, in 2014, the Commission published "Guidelines on State aid for environmental protection and energy 2014–2020" (European Commission 2014) which effectively organised the convergence of RES-E support schemes across Member States. As Solorio and Bocquillon (2017) have noted, these guidelines appear as a new strategy for Europeanising RES-E support schemes, aligning them with the Commission's perspectives on renewable energy policy, and "folding [them] into the framework of the internal electricity market".

The Commission developed such guidelines along with guidance on the design of RES-E support schemes. Ahead of their publication, it presented these guidelines and guidance as a reaction to recent evolutions in RES-E technologies, costs, and policies. It also associated them closely with the achievement of the internal electricity market, as the very titles of the Commission's Communication relevant to RES-policy suggest: "Making the internal market work" (European Commission 2012b); "Renewable energy: a major player in the European energy market" (European Commission 2012a); "Delivering the internal electricity market and making the most of public intervention" (European Commission 2013d). The guidelines were intended "to reflect the technological landscape and EU policy objective in the energy sector, while minimising competition distortion in the internal market" (European Commission 2012b, p. 14) and to ensure that "State aid control facilitates the granting of aid provided that it is well-designed, targeted, least distortive and provided that no better alternative (regulatory, market based instruments) are (sic) available" (European Commission 2012b, p. 14).

Phasing out Feed-in Tariffs

The new guidelines on State aid for environmental protection and energy are not limited to RES-E policy, but they have had important implications concerning the conception of FITs. Along with the Commission's guidance on RES-E support, they amounted to a turn away from FITs, which fell back under the label of "State aid".

Besides integration in the internal electricity market, the Commission stressed three elements to be taken into account in the design and assessment of RES-E policy: the importance of investor's confidence; the limitation of costs for consumers; and the increasing maturity of renewable energy technologies. Regarding the latter, the Commission emphasised that as technologies mature and costs decline, RES-E should move towards full integration in the market and less dependence on State support. It was time for renewables to be exposed to market prices:

> As the renewables sector and technologies mature and grow and as costs decline, it is important that production and investment decisions are driven increasingly by the market and not by guaranteed price levels determined by public authorities. Any support that is still necessary should therefore supplement market prices, not replace them, and be limited to the minimum needed. (European Commission 2013d, p. 15)

Previously, the Commission had already noted that "major negative features" of FITs, including "the impairment of flexible and liquid markets, limiting growth to certain technologies and sizes of installations, and the difficulty in setting appropriate tariff levels and in adjusting such tariffs", "ha[d] been revealed in recent years" (European Commission 2013c, p.11). Unsurprisingly, then, its recommendation was to phase out from feed-in tariffs and to replace them with instruments that "force producers to respond to market prices", such as quota systems or feed-in premiums. These were defended not on grounds of economic efficiency, but of their appropriateness for the market integration of RES-E. The Commission also recommended a turn to tendering procedures, considered as a way "to revel the costs of the different technologies, operators and projects" (European Commission 2013d, p. 16) and as "a self-regulating, subsidy phase out mechanism" (European Commission 2013c).

The Guidelines on State aid effectively organised the phasing out of FITs and their replacement by a combination of feed-in premiums and competitive bidding (with Tradable Green Certificates remaining an option). They mapped out the evolution of requirements, some applying from 2016, others from 2017, with a transitional phase in 2015 and 2016. They were translated in Member States' support schemes: for instance, the revision of the EEG in 2014 and the French Law on Energy Transition for Green Growth both planned the gradual replacement of FITs with a premium system. This seems to suggest that the guidelines were an effective way for the Commission to gain back a degree of control over Member States' renewable energy policies.

Conclusion

In the period between 2008 and 2015, RES-E support schemes were widespread throughout the EU, and many had been several years in operation. FITs seemed to have proved their effectiveness in driving the deployment of RES-E generation. This chapter was interested in their evolution following their widespread validation around 2005. FITs indeed had unexpected effects, and we have explored here how the difficulties in their management led to modifications in their agencement, that is in the discourses about FITs, in the theorisation of FITs, and in the nitty-gritty of policy adjustment.

Tracing the academic and practical issue that emerged, we have shown that FITs—and, with them, RES-E support—were re-politicised, even though the debate about their relevance had seemed settled in the mid-2000s. Debates around FITs were still concerned with the articulation of politics and economic activities and parameters, but the formulation of the problems with this articulation had changed to an extent. The main issues were no longer to internalise externalities, to ensure fair and level competition, or to drive innovation by encouraging niche investment. Two new problems appeared. One was clearly political: the costs of FITs were paid collectively by electricity users, so disproportionate FIT levels compared to technology costs had distributional consequences. The second issue was more market-related, and had to do with investors' confidence: in many countries, FIT schemes were reformed abruptly (notably because of distributional issues), and the ensuing regulatory uncertainty tended to drive investors away. FIT design became increasingly conceived as a matter of compromising between these two problems.

In this process, the consideration for the various markets related to FITs also became more differentiated: the ideal blueprint for the internal electricity market (internalising all external costs and optimising socio-environmental benefits), the actual electricity markets (built around RES-E support as a proxy to internalisation and leading to distributional consequences between consumers and producers), and the markets for renewable energy technologies (growing globally and leading to transfer of support to RES technology between countries).

Another key consequence of the difficult management of FITs after 2008 was the new position of the European Commission. The Commission indeed seized the opportunity to prompt a reconsideration of FITs and eventually reframed FITs as State aid. Instead of pushing for harmonisation, it used a new strategy to assert some form of control over the evolution of RES-E policies by laying out guidelines on State aid that prompted a phasing out of FITs.

NOTES

1. Shum and Watanabe (2008) have shown that experience curve effects driven by production learning and R&D could account for the cost evolution of photovoltaic cells and modules, but that the economics of system integration and deployment in the downstream value chain were driven by more refined inter-project learning.

REFERENCES

Bazilian, Morgan, Ijeoma Onyeji, Michael Liebreich, Ian MacGill, Jennifer Chase, Jigar Shah, Dolf Gielen, Doug Arent, Doug Landfear, and Shi Zhengrong. 2013. Re-considering the economics of photovoltaic power. *Renewable Energy* 53: 329–338. https://doi.org/10.1016/j.renene.2012.11.029.

Cointe, Béatrice. 2017. Managing political market *agencements*: Solar photovoltaic policy in France. *Environmental Politics* 26 (3): 480–501. https://doi.org/10.1080/09644016.2016.1269527.

Commission of the European Communities. 2008. *The support of electricity from renewable energy sources.* Commission staff working document accompanying document to the Proposal for a directive of the European Parliament and of the Council on the promotion of the use of energy from renewable sources [COM(2008) 19 final], SEC(2008) 57, Brussels, 23 January 2008.

Couture, Toby, and Yves Gagnon. 2010. An analysis of feed-in tariffs remuneration models: Implications for renewable energy investment. *Energy Policy* 38: 955–965. https://doi.org/10.1016/j.enpol.2009.10.047.

Dinica, Valentina. 2006. Support systems for the diffusion of renewable energy technologies—An investor perspective. *Energy Policy* 34: 461–480. https://doi.org/10.1016/j.enpol.2004.06.014a.

———. 2008. Initiating a sustained diffusion of wind power: The role of public-private partnerships in Spain. *Energy Policy* 36: 3562–3571. https://doi.org/10.1016/j.enpol.2008.06.008.

European Commission. 2011a. *Renewable energy: progressing towards the 2020 target.* Communication from the Commission to the European Parliament and the Council. COM(2011) 31 final. Brussels, 31 January 2011.

———. 2011b. *Review of European and national financing of renewable energy in accordance with Article 23(7) of Directive 2009/28/CE.* Commission staff working document. Accompanying document to the Communication from the Commission to the European Parliament and Council [COM(2011) 31 final]. SEC(2011) 131 final. Brussels, 31 January 2011.

———. 2012a. *Renewable Energy: A major player in the European energy market.* Communication from the Commission to the European Parliament, the Council, the European Economic and Social Committee and the Committee of the Regions. COM(2012) 271 final, Brussels, 6 June 2012.

———. 2012b. *Making the internal energy market work.* Communication from the Commission to the European Parliament, the Council, the European Economic and Social Committee, and the Committee of the Regions. COM(2012) 663 final, Brussels, 15 November 2012.

———. 2013a. *Renewable energy progress report.* Report from the Commission to the European Parliament, the Council, The European Economic and Social Committee and the Committee of the Regions. COM(2013) 175. Brussels, 27 March 2013.

———. 2013b. *Commission staff working document accompanying the document.* Report from the Commission to the European Parliament, the Council, The European Economic and Social Committee and the Committee of the Regions Renewable energy progress report [COM(2013) 175]. SWD(2013) 102 final. Brussels, 27.03.2013.

———. 2013c. *European Commission guidance for the design of renewables support schemes.* Commission staff working document accompanying the document Communication from the Commission Delivering the internal market in electricity and making the most of public intervention. Draft, October 2013

———. 2013d. *Delivering the internal electricity market and making the most of public intervention.* Communication from the Commission. C(2013) 7243 final, Brussels, 5 November 2013.

———. 2014. Guidelines on State aid for environmental protection and energy 2014–2020. Communication from the Commission. 2014/C 200/01.

Finon, Dominique. 2008. L'inadéquation du mode de subvention du photovolta-ïque à sa maturité technologique. *Working Paper 2008–09,* Paris: CIRED.

Frondel, Manuel, Nolan Ritter, and Christoph M. Schmidt. 2008. Germany's solar cell promotion: Dark clouds on the horizon. *Energy Policy* 36: 4198–4204. https://doi.org/10.1016/j.enpol.2008.07.026.

Frondel, Manuel, Nolan Ritter, Christoph M. Schmidt, and Colin Vance. 2010. Economic impacts from the promotion of renewable energy technologies: The German experience. *Energy Policy* 38: 4048–4056. https://doi.org/10.1016/j. enpol.2010.03.029.

Haas, Reinhaard, Anne Held, Dominique Finon, Niels I. Meyer, Arturo Lorenzoni, Ryan Wiser, and Ken-Ichiro Nishio. 2008. Promoting electricity from renewable energy sources—Lessons learned from the EU, U.S. and Japan. In *Competitive electricity markets,* ed. Fereidoon P. Sioshansi, 91–133. London: Elsevier.

Harmelink, Carlo, Isabelle de Lovinfosse, Michèle Koper, Christina Beestermoeller, Christian Nabe, Matthias Kimmel, Arno van den Bos, Ismail Yildiz and Marieke Harteveld. 2012. *Renewable energy progress and biofuels sustainability. Report for the European Commission.* Report for the European Commission. Tender number ENER/CI/463-2011-Lot2.

Hoppmann, Joern, Joern Huenteler, and Bastien Girod. 2014. Compulsive policy-making—The evolution of the German feed-in tariff system for solar photovoltaic power. *Research Policy* 43 (8): 1422–1441. https://doi.org/10.1016/j. respol.2014.01.014.

Jacobs, David. 2012. *Renewable energy policy convergence in the EU. The evolution of feed-in tariffs in Germany, Spain and France.* Farnham: Ashgate.

Jäger-Waldau, Arnulf. 2013. *PV status report 2013.* JRC Scientific and Technical Reports. Ispra: European Commission Joint Research Centre.

Jamasb, Tooraj. 2007. Technical change theory and learning curves: Patterns of progress in electricity generation technologies. *The Energy Journal* 28 (1): 51–71. https://doi.org/10.5547/ISSN0195-6574-EJ-Vol28-No3-4.

Lesser, Jonathan A., and Xuejuan Su. 2008. Design of an economically efficient feed-in structure for renewable energy development. *Energy Policy* 36: 981–990. https://doi.org/10.1016/j.enpol.2007.11.007.

Loi no 2000-108 du 10 février 2000 relative à la modernisation et au développement du service public de l'électricité. *Journal Officiel de la République Française,* 35, 2143.

Lüthi, Sonja, and Rolf Wüstenhagen. 2012. The price of policy risk—Empirical insights from choice experiments with European photovoltaic project developers. *Energy Economics* 34: 1001–1011. https://doi.org/10.1016/j.eneco. 2011.08.007.

Ménanteau, Philippe, Dominique Finon, and Marie-Laure Lamy. 2003. Prices versus quantities: Choosing policies for promoting the development of renewable energy. *Energy Policy* 31: 799–812. https://doi.org/10.1016/S0301-4215 (02)00133-7.

Mitchell, Catherine, Janet Sawin, Govind R. Pokharel, Daniel Kammen, Zhongying Wang, Solomne Fifita, Mark Jaccard, Ole Langniss, Hugo Lucas, Alain Nadai et al. 2011. Policy, financing and implementation. In *IPCC special report on renewable energy sources and climate change mitigation*, edited by Ottmar Edenhofer, Ramon PichsMadruga, Youba Sokona, Kristin Seyboth, Patrick Matschoss, Susanne Kadner, Timm Zwickel, Patrick Eickemeier, Gerrit Hansen, Steffen Schlömer and Christoph von Stechow. Cambridge and New York, NY: Cambridge University Press.

Ringel, Marc. 2006. Fostering the use of renewable energies in the European Union: The race between feed-in tariffs and green certificates. *Renewable Energy* 31 (1): 1–17.

Schilling, Melissa A. and Melissa Esmundo. 2009. Technology S-curves in renewable energy alternatives: an analysis and implication for industry and government. *Energy Policy* 37: 1767-1789. doi:10.1016/j.enpol.2009.01.004

Schmalensee, Richard. 2012. Evaluating policies to increase electricity generation from renewable energy. *Review of Environmental Economics and Policy* 6: 45–64. https://doi.org/10.1093/reep/rer020.

Shum, Kwok L., and Chihiro Watanabe. 2008. Towards a local learning (innovation) model of solar photovoltaic deployment. *Energy Policy* 26: 508–521. https://doi.org/10.1016/j.enpol.2007.09.015.

Solorio, Israel, and Pierre Bocquillon. 2017. EU renewable energy policy: A brief overview of its history and evolution. In *A guide to EU renewable energy policy: Comparing europeanization and domestic policy change in EU member states*, ed. Israel Solorio and Helge Jörgens. Cheltenham: Edward Elgar Publishing.

van der Zwaan, Bob, and Ari Rabl. 2004. The learning potential of photovoltaics: Implications for energy policy. *Energy Policy* 32: 1545–1554. https://doi. org/10.1016/S0301-4215(03)00126-.

CHAPTER 6

Conclusion

Abstract The concluding chapter summarises the trajectory of feed-in tariffs in the European Union (EU) as retraced in the book, stressing that there are two interwoven storylines: that of the perspective of the European Commission on renewable energy policy, and that which follows the actual, more bottom-up emergence and evolution of feed-in tariffs in Member States. Based upon this brief overview of the main stages in the evolution of feed-in tariffs, Cointe and Nadaï elaborate on a series of crosscutting themes: the close ties between European renewable energy policy and the liberalisation agenda; the ambiguities of the category "market-based" when it describes policy; the diversity of feed-in tariffs agencements; and the frictions between EU-wide and national approaches to renewable energy policy.

Keywords Feed-in tariffs • European Union • Renewable energy • Agencement • Harmonisation

In this book, we have sought to retrace the origins and evolution of feed-in tariffs (FITs) for electricity from renewable energy sources (RES-E) as part of the policy arsenal developed to promote renewable energy in Europe. Following analyses of the construction of the European Union (EU) as a technological, economic and political space around specific objects (Barry 1993, 2001; Laurent 2015; Doganova and Laurent 2016),

© The Author(s) 2018   111
B. Cointe, A. Nadaï, *Feed-in tariffs in the European Union*, Progressive
Energy Policy, https://doi.org/10.1007/978-3-319-76321-7_6

we have argued that the trajectory of FITs could help understand evolving conceptions of renewable energy policy in articulation with the liberalisation and environmental agendas of the EU.

While the adoption of FITs in several EU Member States was initially not directly driven by the EU— the EU Commission first considered it as State intervention in the operation of market and as a potential distortion to competition, it took place under the double umbrella of EU renewable energy policy and of the liberalisation of the EU electricity market. Both the liberalisation agenda and renewable energy policy emerged in Europe in the late 1980s. The former materialised in a Directive on the internal electricity market in 1996, the latter in 2001 with a Directive on the promotion of RES-E. The trajectory of FITs thus needs to be understood in relation to two dimensions of EU energy policy—the promotion of renewable energy and liberalisation—while acknowledging the role and dynamics of FITs development in Member States.

As we emphasised, over this same period, FITs and RES-E policy turned into objects of academic scrutiny. We have tried to take into account this theorisation of FITs and its interplay with EU and (maybe to a lesser extent) national policy-making. In so doing, our objective was to give an account of different manners of making sense of, and of rationalising, the evolution of FITs—that of the European Commission and that of renewable energy policy experts—while also noting the role of the EU in the development of expertise, through the commissioning of assessment reports. Our point in doing so was neither to derive policy recommendations nor to map out extensively the expertise or the Commission's meanders. Rather, it was to analyse the interweaving of theoretical and practical concerns. It was to approach the conception and evolutions of FITs as the joint product of economic expertise and frames of references on the one hand, and of politics and on-the-ground renewable energy policy implementation on the other.

As recent analyses have shown, such a mix is part of the liberalisation agenda and of the construction of the internal market (Reverdy 2014; Doganova and Laurent 2016). It is well captured by the performativity approach in social sciences, which considers theories about the economy as *not* separated from the actual engineering and unfolding of economic activities, and regards the correspondence of economic theories and economic reality as the result of materially enacted constructions involving economic theories as much as practices. Such a commitment is captured under the notion socio-technical agencements—defined as sets of elements put together in view to perform a specific function, which we used here to operationalise our empirical take on FITs and to follow their changes in the EU since their emergence. The notion of agencement conveys the idea of fragility in the articulation of theory and on-the-ground practice, because problems arising from the operation of markets regularly have to be accounted for and addressed through the advancement of knowledge, or the adaptation of on-the-ground devices and practices of market making.

The story about FITs that comes out of this book results from this analytical choice. It is thus an attempt at reading the career of FITs as shifting agencements, early on endowed with the mission to articulate the EU's construction around a single (electricity) market with its political ambition to growth and environmental protection, and shaped at the intersection of economic theories and on-the-ground economic and regulatory practices. This is the reason why this story weaves together, along one timeline, notions from economic theories (visions of ideal markets, categories of costs, notions of risks...), elements from concrete economic descriptions (references to concrete markets, quantities, proxies for valuation of costs, prices ...), and on-the-ground politics and policy-making.

THE TRAJECTORY OF FEED-IN TARIFFS IN EUROPE

Chapter 2 recounted how FITs appeared in the 1970s, with the introduction of devices to integrate wind power into existing electricity systems in Denmark and Germany, where they were gradually turned into renewable energy support instruments. In parallel, the beginnings of an EU renewable energy policy in the mid-1990s were guided by the ideal of a liberalised market as a device for economic and social optimisation. A specific conception of RES-E policy took shape at the European level, which can be traced in the Commission's documents. The position of FITs within this conception is one of the stories we have related in this book.

From the early days of EU energy policy, the Commission understood RES-E policy as a device to articulate environmental objectives, innovation and the liberalisation of the electricity market, that is to achieve a level playing field, a harmonised European economic space through which electricity could flow without restraints. As we described in Chap. 3 (1996–2001), during the elaboration of the 2001 Directive on the promotion of RES-E, the Commission promoted an instrument for RES-E policy that had been designed for this very end: Tradable Green Certificates (TGCs). Back in

2000, TGCs were the only "market-based" solution in the eyes of the Commission: more precisely, TGCs were "internal electricity market-based". The market, though, was invoked as the ideal of a level playing field that would be able, in theory, to account for all social and environmental costs and to make possible a fair competition between fossil and non-fossil energies so as to to steer the realisation of environmental objectives, innovation and economic growth. In contrast, at this time, the Commission considered FITs as State intervention in the market and as subjected to regulatory capture and to the arbitrary shifts of political interests.

However, some Member States were already implementing FITs. They were demonstrating that FITs could successfully sustain the development of RES-E. Partly because of this evidence, and partly because Member States were using them anyway, the Commission had to accept FITs and to postpone the harmonisation of RES-E support that it still hoped to achieve. FITs would then be developed, experimented with and sophisticated, along with alternative support schemes based on the TGC approach. By the late 2000s, FITs seemed to have demonstrated their capacity to articulate the EU's environmental and innovation objectives with the working of market(s).

In this phase (2001–2008), which was the focus of Chap. 4, the European Commission's commitment to coordinate RES-E policy across the Union via regular reviews and assessments encouraged the production of expertise about renewable energy policy instruments. The Commission remained committed to its objective of harmonising RES-E policy across the EU to better insert it in the framework of the internal electricity market, but somewhat softened its stance on FITs. As FITs were re-arranged and sophisticated—in their practice notably with the German *Erneuerbare Energien Gesetz* (EEG), and in theory as research developed—their relationship to so-called market dynamics was gradually recast. In the forms that FITs took by the mid-2005, they appeared able to sustain innovation and drive technology costs down by reducing and remunerating the risk of investing into renewable energy technologies; they could be tailored to specific technologies and levels of technical and economic maturity; they could be made less dependent on politics by indexing their value on installed capacities (as a proxy to the decrease in costs of renewable energy technologies). The sophistication of both the instrument and the expertise also seemed to indicate a shift in focus from competition to investment. At this point, the Commission recognised that FITs were the most efficient type of policy support for RES-E, either implicitly admitting their capacity to fit in the internal electricity market or giving priority to environmental concerns over liberalisation.

In the fifth chapter, we analysed the period after 2008 as a phase of many reforms and reconsiderations in RES-E policies. The very of effects of FITs brought national FIT agencements into new sets of relations, which triggered overflows in various forms. For instance, in the case of photovoltaics, FITs agencements were caught up into transnational market relations. This made the intended support to innovation, growth, and employment in the EU flow to the Chinese photovoltaic industry. The surge in photovoltaic projects that resulted from their high attractiveness for investment also raised the issue of the overall cost of FITs, bringing in the figure of the electricity user, who was paying for RES-E support via a levy on electricity consumption. Such overflows progressively re-opened and re-politicised FITs along new issues such as the distributive effects of risk or industrial policy. These overflows triggered multiple re-adjustments of FITs.

As FITs seemed to create problems and were re-opened and re-discussed politically, the Commission eventually tried to re-qualify them as State aid and to organise their phasing out. It advocated either tenders or premiums that would connect the level of support to prices on the electricity market. The concern for investors' confidence and for the smooth operation of the internal electricity market that the Commission expressed suggested that the internal electricity market was then conceived primarily as a tool for the optimal orientation of investment.

However, the trajectory of FITs cannot be reduced to their career within EU politics and policy. FITs originally appeared before the internal electricity market, and most of their evolution in Member States was not driven by the internal market objective. FITs were arranged in and by Member States, taking shape as diverse agencements to articulate RES-E, environmental objectives, and national electricity markets and systems. Exchanges across countries and evolutions in designs and objectives show varying strategies to manage problems and reforms, and various ways of articulating national RES-E policy within the EU framework: from the shift to TGC in Denmark to anticipate the EU's aborted decision to shift towards pan-European TGC to the resistance to harmonisation led by Germany and Spain who wanted to keep their FIT schemes. Other arguments and objectives than liberalisation, as well as diverse coalitions of interests, have taken part in the evolution in FITs, even though we occasionally find market-based justifications and rhetoric (for instance in Germany) and references to external costs as an invoked benchmark for FIT levels.

LESSONS DRAWN

Renewable Energy Policy and the Internal Electricity Market

The European history of FITs evidences the intimate ties between the EU's renewable energy policy and its liberalisation and harmonisation agendas. The Commission's conception of RES-E support was framed by its political project of achieving Europe-wide integration *via* the market: a frictionless, liberalised market was considered as the instrument for enacting a unified European economic and political space (Barry 1993). By extension, and drawing on textbook economics models of markets, this market was expected to enable the achievement of a range of other objectives of the EU, among which were environmental protection, climate change mitigation, and the promotion of innovation. The promotion of electricity from renewable energy sources was just at the intersection of these objectives: it required technological innovation to become economically competitive; it participated in environmental and climate change policy; and it implied that new players could enter the electricity market, so could foster competition. In the Commission's view, an electricity market that accommodated RES-E (meaning that it tried to mitigate the market imperfections that had hitherto hindered their development) would naturally enhance environmental protection and innovation; at the same time, the establishment of conditions enabling RES-E to enter this market would correct market imperfections that hindered competition and would direct investment towards the "best" options for energy production. This case thus shows how the European liberalisation agenda deploys when applied to a specific concern.

This conception appears to have largely framed the Commission's vision on RES-E, from its push for harmonised European solutions to the references it used to assess Member States policies. It is rather idiosyncratic of the Commission's approach to policy. Its actual influence on Member States policy choices, though hard to evaluate from our material, is likely to have been limited, if only because policy and politics at the national level are shaped by multiple forces and interests. Still, the peculiarities of the Commission's perspective and the history of EU renewable energy policy need to be taken into account to make sense of the trajectory of FITs, even if they cannot explain all of it. The role of the EU in driving the production of expertise on renewable energy policy and in fostering its circulation throughout Europe (via progress reports and

assessment published as Commission documents, for instance) are sugges-
tive of its influence on the course of RES-E policy.

The Ambiguous Relationship Between FITs and the Liberalisation of Electricity Markets

The history of FITs and of how they fit in the liberalisation agenda is not
straightforward. FITs actually appeared before the EU's renewable energy
agenda was fleshed up. They were not initially guided by EU rules in their
design. Our account shows that the European Commission has alternately
framed FITs as undermining the good functioning of the internal electricity market or as contributing to it. This oscillation leads to two more
general reflections.

What Does It Mean to be "Market-Based"?

The case of FITs in the EU invites to interrogate, and open up, the quali-
fication of policy instruments as "market-based". Prior to the 2001
Directive, when the Commission argued for a pan-European TGC scheme,
it described TGCs as "market-based", opposing them to FITs understood
as "regulatory" instruments that risked being considered as State aid
(Lauber and Schenner 2011). Some years later, when the Commission
accepted FITs as the most effective mechanisms for RES-E promotion,
they had become "market-based" in the Commission's accounts (Lauber
and Schenner, 2011). Academic discussions also revealed varying perspectives as to the relationship of FITs with market dynamics (Hvelplund
2001). The delimitation of the "market-based" category is thus not stable:
on the contrary, throughout the process we have described, it was discussed, negotiated, and displaced. This echoes the observation made in
performativity studies that the delimitation of markets and economic
space, and their demarcation from the political, is an outcome of processes
of economisation and marketisation (Callon 2009).

To clarify the somewhat ambiguous reference to markets here, it is
important to note that there have been, in fact, many different forms of
markets at play in our account. First, we have often referred to the
European internal electricity market. This term now designates an entity
that exists: since 1996, the internal electricity market has been partly
implemented, with concrete implications for the organisation of electricity
markets, for their regulation, for pricing techniques, and for electricity
producers and users (Reverdy 2014). The actual internal electricity mar-

ket, however, is quite different from the original blueprint. This is where we encounter a second internal market: the abstract textbook market, infused with conceptions drawn from neoclassical economics, which translates into a political ideology that conceives of perfect markets as not only achievable, but also as an instrument for the common good, able to organise the optimal allocation of resources and investments.

Interestingly, while the Commission regularly invokes the figure of a perfect market (internalising all externalities) as a political ideal, and continuously supports the production of an assessment of external costs, it is the purported imperfections of the actual market (as compared to the ideal) that justify action. Furthermore, the type of action envisioned—that is support to RES-E through FITs—is not of the type that would make an ideal market come true. Following standard economic theory, supporting RES-E technology may at best compensate hidden subsidies to fossil energies and unaccounted externalities: it would add to these imperfections in a view to compensate them and to entice market actors to behave as they would on an ideal market. There is thus a complex relation going on between the abstract textbook market, existing electricity markets, and the internal electricity market in-the-making. To that extent, using Donald MacKenzie's terminology (MacKenzie 2008), the performativity at work is generic rather than effective: in the Commission's discourse, being "market-based" often refers to an articulation to "the market" as a general category rather than to the development of the exact type of market invoked by the theoretical model.

In addition to these two avatars of the internal electricity market, there are multiple national or regional electricity markets at play, each with its own institutions, modalities of pricing, technical and regulatory organisations, and associated industrial, economic, and political interests. The French electricity system, for instance, operates very differently from the Spanish, Nordic or German ones, and electricity is not positioned similarly in the political landscape of these four zones. Though we did not have the space to go into the details of national RES-E support schemes, it is not surprising that FITs took different forms in different countries.

A fourth type of market is relevant to FITs: markets for renewable energy technologies. FITs, as the promotion of RES-E in general, are also meant as incentives to investments and drivers to innovation in renewable energy technologies. The dynamics of the markets for these technologies are thus a crucial factor in the regulation of RES-E support, in particular insofar as they affect the costs of renewable energy projects and, as a consequence, the costs of RES-E generation. These technology markets rarely fit within the boundaries of the EU: for photovoltaics, the massive expansion of cheap photovoltaic panels in China disrupted FIT schemes in Europe, because it made the alignment of support with project costs difficult to maintain, thereby providing opportunities for windfall profits.

The coexistence of many different types and scales of markets relevant to RES-E deployment highlights the ambiguity of the term "market-based", and the difficulties in using it analytically. There are many markets that a policy instrument can be based upon, and in many different ways. Approaching FITs as agencements helps make sense out of the various ways of relating FITs to markets, because it draws attention to how policy concerns are articulated to economic engineering.

Many Agencements of Feed-in Tariffs

Our account shows that FITs are not always the same. Their basic logic is simple and stable, but it leaves room for multiple forms of arrangements (Cointe 2017). FITs have been designed and deployed in many forms throughout their relatively short history. They are always related to some kind of market(s), but in different ways, and with varying rationales: for instance, they can protect from competition on the electricity market while indexing support to evolutions in renewable energy technologies markets. In other words, FITs can be arranged in many ways, and, as a result, can participate in varying articulations of technologies, electricity systems, national politics, and European policy ambitions. From our account, we get a glimpse of the flexibility of FIT agencements in the evolutions in the references used to calibrate FIT rates.

FITs started as a device to incorporate renewable electricity into existing grids and electricity markets. At first, their objective was mainly to make RES-E comparable to conventional electricity. The baseline for setting FIT rates then was the avoided cost of conventional electricity production: RES-E was valued as the amount of conventional electricity that it permitted not to need to generate. The FITs were a device to align the specificities of RES-E producers to those of conventional electric utilities.

When FITs came to be used as an instrument for the promotion of RES-E, the frame of reference changed: external costs became the theoretical reference for FIT levels and for RES-E support more widely. Framing support according to external costs has several implications. First, it implies the recognition that RES-E has benefits that are not accounted for in extant electricity and technology markets: FITs are a device to promote RES-E because RES-E is deemed to have certain desirable qualities that are not valued as they should be. Second, from this, it follows that FITs are meant to correct so-called market failures. This was particularly clear in the Commission's rhetoric in the 1990s: for the Commission, RES-E policy was meant to correct market failures related to environmental and innovation externalities, and should not create additional distortions. The Commission argued that FITs risked distorting competition on the internal market, and so did not favour them.

As RES-E support systems matured, the reference to justify FITs levels was increasingly found in project and technology costs. This reflected increased attention to technology markets and their dynamics, which translated in more sophisticated FIT designs intended to adjust to renewable energy technologies cost evolutions, as with the EEG in Germany. Similar considerations appeared in the literature. This evolution is probably also related to the difficulty to assess and quantify external costs: they can serve as a theoretical justification, but despite the European Commission's continuous support to methodological and empirical research on the evaluation of external costs, these costs are not so easily made operational (Interview, civil servant 1 2012). Project and technology costs appear as a more practical benchmark, even though their evaluation raises its own difficulties. In the fifth chapter, we showed how these difficulties led to discrepancies between the overall costs of FIT policies and the benefits that individual RES-E promoters could reap from FITs. This, in turn, led to consider FIT calibration in a new light, as a balance between collective costs and investors' confidence.

This brief overview of the changes in reference for FIT calibration suggests another evolution in the way the EU approaches markets. In the 1990s and early 2000s, the Commission's focus in describing its ideal market was on competition: "market-based" policies were those that did not distort competition but, on the contrary, facilitated it and made it more fluid. The reference to external costs to an extent reflects this: taking externalities into account is a way to correct the market and to ensure that it is a level playing field, that is to say that new entrants (in this case, RES-E producers) are not disadvantaged. Though competition is still present in the Commission's rhetoric to this day, the focus seems to have shifted towards investment and investors' security. In other words, what is supposed to steer the market towards the common good is not so much level competition, but the optimal allocation of investment. This is a perhaps slight, but noteworthy evolution in the types of economic dynamics that are valued by the Commission. When evaluating RES-E policy options, the Commission no longer seems to consider them as market agencements organising level transactions and competition, but rather as capitalisation agencements organising investment.

European Frictions

One last running theme in our account was the relationship between EU policy and Member States policies. We discussed how the status of FITs has evolved at the EU level and in the literature. The history of FITs also shows frictions between EU-wide ambitions and national policies and markets, and it is perhaps indicative of tensions at play in European energy politics. The history of how FITs relate to EU renewable energy policy reflects evolving strategies for coordinating and articulating European and national agendas.

The Commission's ambition to harmonise RES-E policy across Europe appeared as one driving force in the evolution of European renewable energy policies. To a large extent, it framed the Commission's stance on renewable energy support schemes. All the same, the story we told is that of repeated rebuffs of the Commission's plans for harmonisation. FITs were at the crux of frictions between Member States' policy-making and plans for EU-wide harmonisation: they originated in diverse national pol-icy contexts, as opposed to TGCs, which had been conceived as a device of a harmonised RES-E scheme within the internal electricity market. These frictions prevented the Commission's vision from materialising—its proposals for harmonisation were discarded twice, in no small part due to resistance from FIT-countries.

However, we would argue that we observe a degree of harmonisation, but in a more indirect form. We have described an interplay across EU policy principles, national policy schemes, and scientific expertise, and this interplay was in large part the result of the Commission's approach to policy design and evolution. Given the impossibility to operationalise full-blown harmonisation in the form of a common support scheme, the concern for harmonisation translated into a process of close monitoring and review. The Commission took up a role of coordinator, commission-ing reports on RES-E policy, funding research projects that fed expertise on a range of support instruments, and encouraging the exchange of best practices. It has been argued that this led to some degree of policy conver-gence among EU countries (Jacobs 2012).

When reading documents produced by the Commission, the process is described almost as a scale-one experiment in policy design, under the Commission's control: Member States experiment with a variety of policy strategies, the Commission regularly surveys and reviews them, expertise develops, instruments sophisticate, the Commission diffuses knowledge about what works and what does not, so that in the end, best practices are mapped out and spread throughout the EU. In a way, the formulation of guidelines to influence policy in Member States is an extension of this logic: instead of pushing a fully drawn scheme for harmonisation, the Commission used a somewhat softer approach, outlining good practices and establishing a definition of State aid concerning environmental and energy policy. This seemed to have been more efficient in programming a phase-out of FITs than previous attempts at developing an EU harmonised scheme. Indeed, Member States such as France and Germany have reviewed their RES-E policy legislation to respect these guidelines.

As we have shown, the evolution of RES-E policy has been slightly more chaotic than this discourse suggests. Still, when considering the process and the peculiar articulation of policy assessment and research that characterise it, we do find an influence of the Commission's harmonisation-driven approach. This was, to an extent, part of the agencement of FITs, because this fed the theorisation of their design and a degree of systematisation of techniques to manage the diverse challenges that FITs appeared to raise. The process through which the Commission sought to direct the evolution of RES-E policies and to shape FITs corresponds to a specific organisation of interactions between the EU and its Member States, and this very organisation played a part in the trajectory of FITs.

REFERENCES

Barry, Andrew. 1993. The European Community and European government: Harmonization, mobility and space. *Economy and Society* 22 (3): 314–326.

———. 2001. *Political machines: Governing a technological society*. London: Athlone Press.

Callon, Michel. 2009. Civilizing markets: Carbon trading between *in vitro* and *in vivo* experiments. *Accounting, Organizations and Society* 34: 535–548. https://doi.org/10.1016/j.aos.2008.04.003.

Cointe, Béatrice. 2017. Managing political market *agencements*: Solar photovoltaic policy in France. *Environmental Politics* 26 (3): 480–501.

Doganova, Liliana, and Brice Laurent. 2016. Keeping things different: Coexistence within European markets for cleantech and biofuels. *Journal of Cultural Economy* 9 (2): 141–156.

Hvelplund, Frede. 2001. Political prices or political quantities? A comparison of renewable energy support systems. *New Energy* 5: 18–23.

Jacobs, David. 2012. *Renewable energy policy convergence in the EU. The evolution of feed-in tariffs in Germany, Spain and France.* Farnham: Ashgate.

Lauber, Volkmar, and Elisa Schenner. 2011. The struggle over support schemes for renewable electricity in the European Union: A discursive-institutionalist analysis. *Environmental Politics* 19: 127–141. https://doi.org/10.1080/096 44016.2011.589578.

Laurent, Brice. 2015. The politics of European agencements: Constructing a market of sustainable biofuels. *Environmental Politics* 24: 138–155. https://doi. org/10.1080/09644016.2014.927190.

Mac Kenzie, Donald. 2008. Is economics performative? Option theory and the construction of derivative markets. In *Do Economists make markets? On the performativity of economics,* ed. Donald MacKenzie, Fabian Muniesa, and Lucia Siu, 55–86. Princeton: Princeton University Press.

Reverdy, Thomas. 2014. *La construction politique du prix de l'énergie.* Paris: Presses de Sciences Po.

APPENDIX: LIST OF DOCUMENTS ANALYSED

EUROPEAN INSTITUTIONS

Directives

Directive 96/92/EC of the European Parliament and of the Council of 19 December 1996 concerning common rules for the internal market in electricity. *Official Journal of the European Communities*, L 027, 30/01/1996, 20–29.

Directive 2001/77/EC of the European Parliament and of the Council of 27 September 2001 on the promotion of electricity produced from renewable energy sources in the internal electricity market. *Official Journal of the European Communities*, L 283, 27/10/2001, 33–40.

Directive 28/2009/EC of the European Parliament and of the Council of 23 April 2009 on the promotion of the use of energy from renewable sources and amending and subsequently repelling Directives 2001/77/EC and 2003/30/EC. *Official Journal of the European Union*, L 140, 05/06/2009, 16–62.

Commission

Growth, competitiveness, employment: the challenges and ways forward into the 21st century. White Paper. COM(93) 700. Brussels, December 1993.

For a European Union energy policy. Green Paper. COM(94) 659 final. Brussels, 11.01.1995

© The Author(s) 2018   125
B. Cointe, A. Nadaï, *Feed-in tariffs in the European Union*, Progressive Energy Policy, https://doi.org/10.1007/978-3-319-76321-7

An energy policy for the European Union. White Paper. COM(95) 682 final. Brussels, 13.12.1995.

Energy for the future: renewable sources of energy. Green Paper for a Community strategy. Communication from the Commission. COM(96) 576 final. Brussels, 20.11.1996.

Energy for the future: renewable sources of energy. White Paper for a Community Strategy and Action Plan. COM(97) 599 final. Brussels, 26.11.1997.

Commission report to the Council and the European Parliament on Harmonization requirements. Directive 96/92 concerning rules for the internal market in electricity. COM(1998) 167, Brussels, 16.03.1998.

Electricity from renewable energy sources and the internal electricity market. Commission working document. SEC(1999) 470 final, Brussels, 13.04.1999.

External Costs: Research results on socio-environmental damages due to electricity and transport. DG Research Report: EUR 20198. Brussels, 2003.

The share of renewable energy in the EU. Commission report in accordance with Article 3 of Directive 2001/77/EC, evaluation of the effect of legislative instruments and other Community policies on the development of the contribution of renewable energy sources in the EU and proposals for concrete action. COM(2004) 366 final. Brussels, 26.05.2004.

The support of electricity from renewable sources. Communication from the European Commission. COM(2005) 627 final. Brussels, 07.12.2005.

A European strategy for sustainable, competitive and secure energy. Green Paper. COM(2006) 105 final. Brussels, 08.03.2006.

Towards a European Strategic Energy Technologies Plan. Communication from the Commission to the Council, the European Parliament, the European Economic and Social Committee and the Committee of the Regions. COM(2006) 847 final. Brussels, 10.01.2007.

Renewable energy roadmap. Renewable energies in the 21st century: building a more sustainable future. Communication from the Commission to the Council and the European Parliament. COM(2006) 848 final. Brussels, 10.01.2007.

An Energy Policy for Europe. Communication from the Commission to the European Council and the European Parliament. COM(2007) 1 final. Brussels, 10.01.2007.

Limiting global climate change to 2 degrees Celsius. The way ahead for 2020 and beyond. Communication from the Commission to the Council, the

European Parliament, the European Economic and Social Committee and the Committee of the Regions. COM(2007) 2 final. Brussels, 10.01.2007.

The support of electricity from renewable energy sources. Commission staff working document accompanying document to the Proposal for a directive of the European Parliament and of the Council on the promotion of the use of energy from renewable sources [COM(2008) 19 final], SEC(2008) 57, Brussels, 23.01.2008.

Renewable energy: progressing towards the 2020 target. Communication from the Commission to the European Parliament and the Council. COM(2011) 31 final. Brussels, 31.01.2011.

Review of European and national financing of renewable energy in accordance with Article 23(7) of Directive 2009/28/CE. Commission staff working document. Accompanying document to the Communication from the Commission to the European Parliament and Council [COM(2011) 31 final]. SEC(2011) 131 final. Brussels, 31.01.2011.

Renewable Energy: a major player in the European energy market. Communication from the Commission to the European Parliament, the Council, the European Economic and Social Committee and the Committee of the Regions. COM(2012) 271 final, Brussels, 06/06/2012.

Making the internal energy market work. Communication from the Commission to the European Parliament, the Council, the European Economic and Social Committee, and the Committee of the Regions. COM(2012) 663 final, Brussels, 15/11/2012.

Renewable energy progress report. Report from the Commission to the European Parliament, the Council, the European Economic and Social Committee and the Committee of the Regions. COM(2013) 175. Brussels, 27.03.2013.

Commission staff working document accompanying the document Report from the Commission to the European Parliament, the Council, The European Economic and Social Committee and the Committee of the Regions renewable energy progress report [COM(2013) 175]. SWD(2013) 102 final. Brussels, 27.03.2013.

European Commission guidance for the design of renewables support schemes. Commission staff working document accompanying the document Communication from the Commission Delivering the internal market in electricity and making the most of public intervention. Draft, October 2013.

Delivering the internal electricity market and making the most of public intervention. Communication from the Commission. C(2013) 7243 final, Brussels, 05/11/2013.

Guidelines on State aid for environmental protection and energy 2014–2020. Communication from the Commission. 2014/C 200/01.

A framework strategy for a resilient energy union with a forward-looking climate change policy. Communication from the Commission to the European Parliament, the Council, the European Economic and Social Committee, the Committee of the Regions and the European Investment Bank. Energy Union Package. COM(2015) 80 final. Brussels, 25/02/2015.

Others

Council resolution of 16 September 1986 concerning new Community energy policy objectives for 1995 and convergence of the policies of the Member States. *Official Journal of the European Communities*, C 241, 25/09/1986, 1–3.

Council recommendation of 9 June 1988 on developing the exploitation of renewable energy sources in the Community. *Official Journal of the European Communities*, L 160, 28/06/1988, 46–48.

European Parliament. 1996. *Mombaur own-initiative report on a community action plan for renewable energy sources*, OJ C 211. http://eur-lex.europa.eu/legal-content/EN/TXT/PDF/?uri=OJ:C:1996:211:FUL L&from=EN

European Parliament. 1998. *Report on network access for renewable energies—creating a European directive on the feeding in of electricity from renewable sources of energy in the European Union; rapporteur Linkohr.* A4-0199 and 98, PE224.949.fin, 26 May.

European Court of Justice. 2000. *Opinion of advocate general Jacobs delivered on 26 October 2000*, Case C-379/98 PreussenElektra v. Schleswag.

European Court of Justice. 2001. *PreussenElektra v. Schleswag.* Judgement of the Court, 13 March 2001. Case C-379/98.

Reports and Grey Literature

Coenraads, Rogier, Gemma Reece, Monique Voogt, Mario Ragwitz, Anne Held, Gustav Resch, Thomas Faber, Reinhard Haas, Inga Konstantinaviciute, Juraj Krivošík and Tomas Chadim. 2008.

PROGRESS: promotion and growth of renewable energy sources and systems. Final report. Utrecht, 5 March 2008.

Couture, Toby, Karlynn Cory, Claire Kreycik and Emily Williams. 2010. *A Policymaker's guide to feed-in tariff policy design.* Golden, CO: NREL.

de Jager, David, Corinna Klessmann, Eva Stricker, Thomas Winkel, Erika de Visser, Michèle Koper, Mario Ragwitz, Anne Held, Gustav Resch, Sebastian Busch, et al. 2011. *Financing renewable energy in the European energy market.* Project N° PECPNL084659. Ecofys/DG ENER. 2 January 2011.

Harmelink, Carlo, Isabelle de Lovinfosse, Michèle Koper, Christina Beestermoeller, Christian Nabe, Matthias Kimmel, Arno van den Bos, Ismail Yildiz and Marieke Harteveld. 2012. *Renewable energy progress and biofuels sustainability. Report for the European Commission.* Report for the European Commission. Tender number ENER/CI/463-2011-Lot2.

Ecorys. 2010. *Assessment of non-cost barriers to renewable energy growth in EU Member States—AEON.* Final report. Rotterdam, 10 May 2010.

Held, Anne, Mario Ragwitz, Claus Huber, Gustav Resch, Thomas Faber and Katarina Vertin. 2007. *Feed-in systems in Germany, Spain and Slovenia: a comparison.* Karlsruhe, Germany, October 2007.

Klein, Arne, Erik Merkel, Benjamin Pfluger, Anne Held, Mario Ragwitz, Gustav Resch and Sebastian Busch. 2010. *Evaluation of different feed-in tariff design options—Best practice paper for the International Feed-in Cooperation.* 3nd edition. Fraunhofer Institute.

Peters, Daan, Sacha Alberici, Gemma Toop and Bettina Kretschmer. 2012. *Analysis of Member States RED implementation. Final report.* Utrecht, 13 December 2012.

Ragwitz, Mario, Anne Held, Gustav Resch, Thomas Faber, Reinhard Haas, Claus Huber, Poul Erik Morthorst, Stine Grenaa Jensen, Rogier Coenraads, Monique Voogt, Gemma Reece, Inga Konstantinaviciute and Bernhard Heyder. 2007. *Assessment and optimisation of renewable energy support schemes in the European electricity market. OPTRES final report.* Karlsruhe, February 2007.

Ragwitz, Mario, Jenny Winkler, Corinna Klessmann and Gustav Resch. 2012. *Recent experiences with FIT systems in the European Union—a research paper for the International feed-in cooperation.* BMU commissioned report.

Academic Papers

Agnolucci, Paolo. 2007a. The importance and the policy impacts of post-contractual opportunism and competition in the English and Welsh non-fossil fuel obligation. *Energy Policy* 35: 475–486.

Agnolucci, Paolo. 2007b. The effect of financial constraints, technological progress and long-term contracts on tradable green certificates. *Energy Policy* 35: 3347–3359.

Anderson, Philip and Michael L. Tushman. 1990. Technological discontinuities and dominant designs: a cyclical model of technological change. *Administrative Science Quarterly* 35(4): 604–633.

Anthoff, David and Robert Hahn. 2010. Government failure and market failure: on the inefficiency of environmental and energy policy. *Oxford Review of Economic Policy* 26(2): 197–224.

Awerbuch, Shimon. 2000. Investing in photovoltaic: risk, accounting and the value of new technology. *Energy Policy* 28: 1023–1035.

Badcock, Jeremy and Manfred Lenzen. 2010. Subsidies for electricity-generating technologies: a review. *Energy Policy* 38: 5038–5047.

Barradale, Merrill Jones. 2010. Impact of public policy uncertainty on renewable energy investment: wind power and the production tax credit. *Energy Policy* 38: 7698–7709.

Bergek, Anna and Staffan Jacobsson. 2010. Are tradable green certificates a cost-efficient policy driving technical change or a rent-generating machine? Lessons from Sweden 2003–2008. *Energy Policy* 38: 1255–1271.

Bergek, Anna, Staffan Jacobsson, Bo Carlsson Sven Lindmark and Annika Rickne. 2008. Analyzing the functional dynamics of technological innovation systems: a scheme of analysis. *Research Policy* 37: 407–429.

Bergek, Anna, Staffan Jacobsson and Björn A. Sandén. 2008. 'Legitimation' and 'development of positive externalities': two key processes in the formation phase of technological innovation systems. *Technology Analysis & Strategic Management* 20(5): 575–592.

Blok, Kornelius. 2006. Renewable energy policies in the European Union. *Energy Policy* 34: 251–255.

Bürer, Mary Jean and Rolf Wüstenhagen. 2009. Which renewable energy policy is a venture capitalist's best friend? Empirical evidence from a survey of international cleantech investors. *Energy Policy* 37: 4997–5006.

Büsgen, Uwe and Wolfhart Dürrschmidt. 2009. The expansion of electricity generation from renewable energies in Germany: A review based on the Renewable Energy Sources Act Progress Report 2007 and the new German feed-in legislation. *Energy Policy* 37: 2536–2545.

Butler, Lucy and Karsten Neuhoff. 2008. Comparison of feed-in tariff, quota and auction mechanisms to support wind power development. *Renewable Energy* 33: 1854–1867.

Carlsson, Bo and Rickard Stankiewicz. 1991. On the nature, function and composition of technological systems. *Journal of Evolutionary Economics* 1: 93–118.

Couture, Toby and Yves Gagnon. 2010. An analysis of feed-in tariffs remuneration models: implications for renewable energy investment. *Energy Policy* 38: 955–965.

Dewald, Ulrich and Bernhard Truffer. 2011. Market formation in technological innovation systems—Diffusion of photovoltaic applications in Germany. *Industry and Innovation* 18: 285–300.

Dinica, Valentina. 2006. Support systems for the diffusion of renewable energy technologies—an investor perspective. *Energy Policy* 34: 461–480.

Dinica, Valentina 2008. Initiating a sustained diffusion of wind power: the role of public-private partnerships in Spain. *Energy Policy* 36: 3562–3571.

Dinica, Valentina 2011. Renewable electricity production costs: a framework to assist policy-makers decisions on price support. *Energy Policy* 39: 4153–4167.

Espey, Simone. 2001. Renewable portfolio standards: a means for trade with electricity from renewable energy sources? *Energy Policy* 29: 557–566.

Finon, Dominique and Philippe Ménanteau. 2004. The static and dynamic efficiency of instruments of promotion of renewables. *Energy Studies Review* 12: 53–83.

Finon, Dominique. 2008. L'inadéquation du mode de subvention du photovoltaïque à sa maturité technologique. *Working Paper* 2008–09, Paris: CIRED.

Fischer, C. and Preonas, L. (2010). Combining policies for renewable energy: is the whole less than the sum of its parts? *International Review of Environmental and Resource Economics* 4, 51–92.

Fischer, Caroly and Richard G. Newell. 2008. Environmental and technology policies for climate mitigation. *Journal of Environmental Economics and Management* 55: 142–162.

Fouquet, Doerte and Thomas B. Johansson. 2008. European renewable energy policy at crossroads—Focus on electricity support mechanisms. *Energy Policy* 36: 4079–4092.

Foxon, Tim and Peter Pearson. 2008. Overcoming barriers to innovation and diffusion of cleaner technologies: some features of a sustainable innovation policy regime. *Journal of Cleaner Production* 16S1: S148-S161.

Freeman, Chris. 1996. The greening of technology and models of innovation. *Technological Forecasting and Social Change* 53: 27–39.

Frondel, Manuel, Nolan Ritter and Christoph M. Schmidt. 2008. Germany's solar cell promotion: dark clouds on the horizon. *Energy Policy* 36: 4198–4204.

Frondel, Manuel, Nolan Ritter, Christoph M. Schmidt and Colin Vance. 2010. Economic impacts from the promotion of renewable energy technologies: the German experience. *Energy Policy* 38: 4048–4056.

Grubb, Michael 2004. Technology innovation and climate change policy: an overview of issues and options. *Keio Economic Studies* 41(2): 103–132.

Haas, Reinhaard, Wolfgang Eichhammer, Claus Huber, Ole Langniss, Arturo Lorenzoni, Reinhard Madlener, Philippe Ménanteau, Poul Erik Morthorst, Alvaro Martins, Anna Oniszk et al. 2004. How to promote renewable energy systems successfully and effectively. *Energy Policy* 32: 833–839.

Haas, Reinhaard, Anne Held, Dominique Finon, Niels I. Meyer, Arturo Lorenzoni, Ryan Wiser and Ken-Ichiro Nidhio (2007). "Promoting electricity from renewable energy sources—lessons learned from the EU, U.S. and Japan". In *Competitive electricity markets,* edited by Fereidoon P. Sioshansi. London: Elsevier, 91–133.

Haas, Reinhaard, Gustav Resch, Christian Panzer, Sebastian Busch, Mario Ragwitz and Anne Held. 2011. Efficiency and effectiveness of promotion systems for electricity generation from renewable energy sources— Lessons from EU countries. *Energy* 36, 2186–2193.

Harmelink, Mirjam, Monique Voogt and Clemens Cremer. 2006. Analysing the effectiveness of renewable energy supporting policies in the European Union. *Energy Policy* 34: 343–351.

Hoppmann, Joern, Joern Huenteler and Bastien Girod. 2014. Compulsive policy-making—The evolution of the German feed-in tariff system for solar photovoltaic power. *Research Policy* 43(8): 1422–1441.

Jacobs, David 2010. Fabulous feed-in tariffs. *Renewable Energy Focus,* July-August 2010, p. 28–30.

Jacobs, David. 2012. *Renewable energy policy convergence in the EU. The evolution of feed-in tariffs in Germany, Spain and France.* Farnham, UK; Ashgate?

Jacobsson, Staffan and Anna Johnson. 2000. The diffusion of renewable energy technology: an analytical framework and key issues for research. *Energy Policy* 28: 625–640.

Jacobsson, Staffan and Volkmar Lauber. 2006. The politics and policy of energy system transformation—explaining the German diffusion of renewable energy technology. *Energy Policy* 34: 256–276.

Jacobsson, Staffan, Anne Bergek, Dominique Finon, Volkmar Lauber, Catherine Mitchell, David Toke and Ariel Verbruggen. 2009. EU renewable energy support: faith or facts? *Energy Policy* 37: 2143–2146.

Jacobsson, Staffan, Björn Sandén and Lennart Bangens. 2004. Transforming the energy system—the evolution of the German technological system for solar cells. *Technology Analysis and Strategic Management* 16: 3–30.

Jaffe, Adam B. and Robert N. Stavins. 1995. Dynamic incentives of environmental regulations: the effects of alternative policy instruments on technology diffusion. *Journal of Environmental Economics and Management* 29: S43-S63.

Jamasb, Tooraj. 2007. Technical change theory and learning curves: patterns of progress in electricity generation technologies. *The Energy Journal* 28(1): 51–71.

Jänicke, Martin and Stefan Lindemann. 2010. Governing environmental innovations. *Environmental Politics* 19(1): 127–141.

Johnstone, Nick, Ivan Hascic and David Popp. 2010. Renewable energy policies and technological innovation: evidence based on patent counts. *Environmental Resources Economics* 45: 133–155.

Jordan, Andrew and Andrea Lenschow. 2000. 'Greening' the European Union: what can be learned from the 'leaders' of EU environmental policy? *European Environment* 10: 109–120.

Klessmann, Corinna, Christian Nabe and Karsten Burges. 2008. Pros and cons of exposing renewables to electricity market risks—a comparison of the market integration approaches in Germany, Spain, and the UK. *Energy Policy* 36: 3646–3661.

Lauber, Volkmar and Lutz Mez. 2004. Three decades of renewable electricity policies in Germany. *Energy and Environment* 15: 599–623.

Lauber, Volkmar and Elisa Schenner. 2011. The struggle over support schemes for renewable electricity in the European Union: a discursive-institutionalist analysis. *Environmental Politics* 19: 127–141.

Lauber, Volkmar. 2004. REFIT and RPS: options for a harmonised Community framework. *Energy Policy* 32: 1405–1414.

Lecuyer, Oskar and Philippe Quirion. 2013. Can uncertainty justify overlapping policy instruments to mitigate emissions? *Ecological Economics.* 93: 177–191.

Lesser, Jonathan A. and Xuejuan Su. 2008. Design of an economically efficient feed-in structure for renewable energy development. *Energy Policy* 36: 981–990.

Lipp, Judith. 2007. Lessons for effective renewable energy policies from Denmark, Germany and the United Kingdom. *Energy Policy* 35: 5481–5495.

Loiter, Jeffrey M.and Vicki Norberg-Bohm. 1999. Technology policy and renewable energy: public roles in the development of new energy technologies. *Energy Policy* 27: 85–97.

Lund, P. D. 2009. Effects of energy policies on industry expansion in renewable energy. *Renewable Energy* 34: 53–64.

Lüthi, Sonja and Rolf Wüstenhagen. 2012. The price of policy risk—Empirical insights from choice experiments with European photovoltaic project developers. *Energy Economics* 34: 1001–1011.

McDonald, Alan and Lea Schrattenholzer. 2001. Learning rates for energy technologies. *Energy Policy* 29: 255–261.

Ménanteau, Philippe, Dominique Finon and Marie-Laure Lamy. 2003. Prices versus quantities: choosing policies for promoting the development of renewable energy. *Energy Policy* 31: 799–812.

Meyer, Niels I. 1995. Danish wind power development. *Energy for Sustainable Development* 2: 18–25.

Meyer, Niels I. 2003. European schemes for promoting renewables in liberalised markets. *Energy Policy* 31: 665–676.

Meyer, Niels I. 2004. Renewable energy policy in Denmark. *Energy for Sustainable Development* 8: 25–35.

Midttun, Atle and Kristian Gautesen. 2007. Feed in or certificate, competition or complementarity? Combining a static efficiency and a dynamic innovation perspective on the greening of the energy industry. *Energy Policy* 35: 1419–1422.

Midttun, Atle and Svein Kamfjord. 1999. Energy and environmental governance under ecological modernization: a comparative analysis of Nordic countries. *Public Administration* 77: 873–895.

Midttun, Atle and Anne Louise Koefoed. 2003. Greening of electricity in Europe: challenges and developments. *Energy Policy* 31: 677–687.

Midttun, Atle. 2003. Introduction: Green electricity in Europe—environment, politics and markets. *Energy Policy* 31: 579–581.

Mitchell, C., Bauknecht, D. & Connor, P. M. (2006). Effectiveness through risk reduction: a comparison of the renewable obligation in England and Wales and the feed-in system in Germany. *Energy Policy,* 34, 297–305.

Mitchell, Catherine, Janet Sawin, Govind R. Pokharel, Daniel Kammen, Zhongying Wang, Solomne Fifita, Mark Jaccard, Ole Langniss, Hugo Lucas, Alain Nadai et al. 2011. Policy, financing and implementation. In *IPCC Special Report on Renewable Energy Sources and Climate Change Mitigation,* edited by Ottmar Edenhofer, Ramon PichsMadruga, Youba Sokona, Kristin Seyboth, Patrick Matschoss, Susanne Kadner, Timm Zwickel, Patrick Eickemeier, Gerrit Hansen, Steffen Schlömer and Christoph von Stechow. Cambridge, United Kingdom and New York, NY, USA: Cambridge University Press.

Nemet, Gregory F. 2006. Beyond the learning curve: factors influencing cost reductions in photovoltaics. *Energy Policy* 34: 3218–3232.

Norberg-Bohm, Vicki. 1999. Stimulating green technological innovation: an analysis of alternative policy mechanisms. *Policy Sciences* 32: 13–38.

Papineau, Maya. 2006. An economic perspective on learning curves and dynamic economies in renewable energy technologies. *Energy Policy* 34: 422–432.

Ringel, Marc. 2006. Fostering the use of renewable energies in the European Union: the race between feed-in tariffs and green certificates. *Renewable Energy* 31: 1–17.

Sandén, Björn A. and Christian Azar. 2005. Near-term technology policies or long-term climate targets? Economy-wide versus technology-specific approaches. *Energy Policy* 33: 1557–1576.

Schilling, Melissa A. and Melissa Esmundo. 2009. Technology S-curves in renewable energy alternatives: an analysis and implication for industry and government. *Energy Policy* 37: 1767–1789.

Schmalensee, Richard. 2012. Evaluating policies to increase electricity generation from renewable energy. *Review of Environmental Economics and Policy* 6: 45–64.

Shum, Kwok L. and Chihiro Watanabe. 2008. Towards a local learning (innovation) model of solar photovoltaic deployment. *Energy Policy* 26: 508–521.

Shum, Kwok L. and Chihiro Watanabe. 2007. Photovoltaic deployment strategy in Japan and the USA—an institutional appraisal. *Energy Policy* 35: 1186–1195.

Solangi, K. H., M. R. Islam, R. Saidur, N. A. Rahim and H. Fayaz, H. 2011. A review on global solar energy policy. *Renewable and Sustainable Energy Reviews*, 15, 2149–2163.

Stenzel, Till and Alexander Frenzel. 2008. Regulating technological change—the strategic reactions of utility companies towards subsidy policies in the German, Spanish and UK electricity markets. *Energy Policy* 36: 2645–2657.

Timilsina, Govinda R., Lado Kurdgelashvili and Patrick A. Narbel. 2012. Solar energy: markets, economics and policies. *Renewable and Sustainable Energy Reviews* 16: 449–465.

Toke, David and Volkmar Lauber. 2007. Anglo-Saxon and German approaches to neoliberalism and environmental policy: The case of financing renewable energy. *Geoforum* 38: 677–687.

Toke, David. 2007. Renewable financial support systems and cots-effectiveness. *Journal of Cleaner Production* 15: 280–287.

Tsoutos, Theocharis D. and Yeoryios Stamboulis. 2005. The sustainable diffusion of renewable energy technologies as an example of innovation-focused policy. *Technovation* 25: 753–561.

Unruh, Gregory C. 2000. Understanding carbon lock-in. *Energy Policy* 28: 817–830.

Unruh, Gregory C. 2002. Escaping carbon lock-in. *Energy Policy* 30: 317–325.

van der Zwaan, Bob and Ari Rabl. 2004. The learning potential of photo-voltaics: implications for energy policy. *Energy Policy* 32: 1545–1554.

Watanabe, Chihiro, Kouji Wakabayashi and Toshinori Miyazawa. 2000. Industrial dynamism and the creation of a "virtuous cycle" between R&D, market growth and price reduction. The case of photovoltaic power generation (PV) development in Japan. *Technovation* 20: 299–312

Weitzman, Martin L. 1974. Price vs. quantities. *The Review of Economic Studies* 41: 477–491.

Weyant, John P. 2011. Accelerating the development and diffusion of new energy technologies: beyond the "valley of death". *Energy Economics* 33: 674–682.

Wiser, Ryan H. and Steven J. Pickle. 1998. Financing investments in renewable energy: the impacts of policy design. *Renewable and Sustainable Energy Reviews* 2: 361–386.

Index[1]

[1] Note: Page numbers followed by 'n' refer to notes.